Mokele-Mbembe

WILLIAM REBSAMEN

Mokele-Mbembe

Mystery Beast of the Congo Basin

William J. Gibbons

Coachwhip Publications
Landisville, Pennsylvania

Mokele-Mbembe: Mystery Beast of the Congo Basin
Copyright © 2010 William J. Gibbons

Cover design by William Rebsamen

ISBN 1-61646-010-5
ISBN-13 978-1-61646-010-5

CoachwhipBooks.com

All Rights Reserved. No part of this publication may be reproduced, stored in a retrieval system or transmitted in any form or by any means—electronic, mechanical, photocopy, recording or any other—except for brief quotations in printed reviews, without the prior permission of the author or publisher.

Contents

Introduction	9
Dragons or Dinosaurs?	11
The Dawn of Adventure	17
The Dragon Hunters	27
Tales from the Heart of Darkness	33
The Expeditions: 1979 to 1992	47
Target–Cameroon!	109
Adventures in the Forbidden Zone	149
Return to the Forbidden Zone	175
Milt Marcy Ventures Forth	187
The MonsterQuest Hunt	191
Mokele-mbembe–A Living Dinosaur?	205
N'Goubou–The Horned Terror	233
Yoli—The Snake Dragon	249
Appendix I: Lingala Words and Phrases	259
Appendix II: Baka Words and Phrases	261
Appendix III: Ethnological Index of Mystery Animal Names	265
Bibliography	267
In Memoriam	269

William Rebsamen

Acknowledgments

I wish to thank the following individual for their support, friendship and invaluable assistance in my search for *mokele-mbembe* over many years:

Roy P. Mackal, Karl Shuker, Loren Coleman, Jo Sarsby, John Kirk III, Chad Arment, Rob Mullin, Milt Marcy, Peter Beach, Brian Sass, William Cooper, Kent Hovind, David Woetzel, Pierre Sima Noutchegeni, Paul Rockel, Paul & Diane Ohlin, Sarah Speer, Pastor Matena Paul, Claude & Jennifer Daouste, Phil & Reda Anderton, Minnie Stoumbaugh, Greg Richardson, Richard Syrett, Michael Coren, Clint Kelly, Bill Rebsamen, Jerry Akridge, Laurence Tisdall, Garth Guessman, Fred L. Light, Ed Holroyd, Don Chittick, Dennis Swift, Dennis Peterson, Ian Taylor, Tom Hall, Pastor Mowawa Eugene, Eugene & Sandy Thomas, Mark Rothermel, Jonathan Walls, Joe Della Porta, Elizabeth Addy, John & Gloria Wilson, and Don & Betty Hocking.

Special thanks also goes to:
The Government of the Republic of the Congo
The Government of the Republic of Cameroon

COURTESY DR. ROY P. MACKAL / *A Living Dinosaur?*

The New York Herald, for February 13, 1910.

Introduction

The very idea of dinosaurs—those "terrible lizards"—still living in some remote corner of the world today threatens to evoke sidesplitting laughter from the modern paleontologist. Surely no scientist worth his salt would even entertain such a preposterous idea. Yet, there are a few dissenting voices within the scientific community, although often marginalized, that beg to differ—and for good reason.

According to conventional evolutionary wisdom, the last of the dinosaurs became extinct over 65 million years ago during a period known as "the great dying." Modern science entertains about thirty different hypotheses on the reason for the extinction. Currently, the most popular theory postulates that the earth was hit by a gigantic meteorite or comet that hit the earth with unprecedented force, sending a shock wave around the earth while simultaneously throwing up a cloud of dust that blocked the life-giving rays of the sun. Thus, the plants that relied on photosynthesis soon died, denying the only food source available to herbivorous dinosaurs, which in turn removed a ready supply of fresh meat for the carnivorous monsters that preyed upon them.

So teach the colorful books and science publications that have shaped the thinking of the generations since Darwin. However, is this really the case?

Throughout history, over 200 cultures around the world have left us with a rich and detailed record of monstrous creatures that they knew as "dragons." These are not the fanciful creatures of familiar fairy tales, but monstrous reptiles that were, for the most part, a menace to those communities and towns that had to deal with them.

From such evidence as the Behemoth of Job 41 to the historical records of 16th century England, and present-day eyewitness accounts, I propose that dinosaurs, or at least creatures that look remarkably like them, are still very much with us today in the dark and remote recesses of our modern, fast-paced world.

So, let us begin our journey together and track those Terrible Lizards!

> Behold now behemoth,
> which I made with thee.
>
> —Job 40:15

1

DRAGONS OR DINOSAURS?—ANCIENT RECORDS

THE BIBLE

Our journey starts within the pages of oldest book in the world—Job. The story of Job tells of a righteous man who lost his livestock, servants, and children in one day, leaving him destitute and alone. His wife admonished him to "curse God and die," which Job refused to do. Eventually, the Lord takes Job and shows him a remarkable creature, the behemoth.

> "Behold now behemoth, which I made with thee; he eateth grass as an ox. Lo now, his strength is in his loins and his force is in the navel of his belly. He moveth his tail like a cedar: the sinews of his stones are wrapped together. His bones are as strong pieces of brass; his bones are like bars of iron. He is the chief of the ways of God: he that made him can make his sword to approach unto him. Surely the mountains bring him forth food, where all the beasts of the field play. He lieth under the shady trees, in the covert of the reed, and fens. The shady trees cover him with their shadow; the willows of the brook compass him about. Behold, he drinketh up a river, and hasteth not: he trusteth that he can draw up Jordan into his mouth. He taketh it with his eyes: his nose pierceth through snares." *JOB 40:15-24*

Bible scholars have suggested the behemoth was an elephant, a hippopotamus, or even an ox, but none of these animals fit the description of the behemoth satisfactorily. The narrative describes behemoth as "chief of the works of God." In other words, the behemoth is the largest and most powerful animal ever created. The biggest of all land animals known to have lived were the mighty dinosaurs. Elephants and hippos do not possess tails like a "cedar," a tree that in biblical literature is illustrative of height and strength. The behemoth, as described in Job, is a much more extraordinary creature than any known living animal. Rather, it can be argued that such a description is reminiscent of a giant sauropod (long-necked) dinosaur such as the *Diplodocus*. In fact, those who assert that behemoth was an elephant or hippopotamus are only following speculations from a period before dinosaurs were even known to have existed.

The behemoth is not the only dinosaur or dinosaur-like animal mentioned in the Bible. Indeed, the Old Testament mentions dragons thirty-five times, and virtually all nations of antiquity seem to have had a similar obsession with dinosaur-like animals they called "dragons." The Biblical writers, inspired by the Holy Spirit, and perhaps as first-hand eyewitnesses themselves, also wrote about dragons. The first reference to created animals, in Genesis 1:21, says "God created great whales," but the Hebrew word translated here as "whales" (*tanniyn*) is translated "dragons" in over 20 other passages. Many of these dragons were sea dwellers, such as the leviathan (Job 41:1-34), and others were flying monsters (Isaiah 30:6).

THE MIDDLE EAST

In 600 B.C., under the reign of King Nebuchadnezzar, a Babylonian artist was commissioned to shape reliefs of animals on the structures associated with the Ishtar Gate. Centuries later, on June 3, 1887 A.D., German archaeologist Robert Koldewey stumbled upon the remains of a blue-glazed brick, and the gate was rediscovered. The animals appear in alternating rows with lions, fierce bulls (*rimi* or *reems* in Chaldean), and curious long-necked dragons (*sirrush*). The lions and bulls would have been present at that time in the Middle East. But on what creature did the ancient Babylonians model the dragon? The same word, *sirrush*, is mentioned in the story of Bel and the Dragon, from the Apocryphal book of Daniel. Both the description there and the image on these unearthed walls, which are now displayed in the Berlin Vorderasiatisches Museum, appear to represent a somewhat stylized dinosaur, including clawed toes, a slender neck, and a single "horn." This book of the Apocrypha tells a curious tale of a dragon that was kept in the temple of the pagan god, Bel, by King Nebuchadnezzar. The people were ordered to worship this great dragon, but the prophet Daniel refused. The king admonished Daniel to worship the monster, reasoning that as it was able to eat and drink, it was a living god, unlike the pagan deities fashioned from brass and iron.

> "And in that place was a great dragon or serpent, which they of Babylon worshipped. And the king said unto Daniel, Wilt thou also say that this is of brass? Lo, he liveth, he eateth and drinketh; thou

canst not say that he is no living god: therefore worship him. Then said Daniel. I will worship the Lord my God: for He is the living God. But give me leave O king, and I shall slay this dragon without sword or staff. The king said, I give thee leave."

Daniel then made several large "cakes" from hair, bitumen, and fat, which he then forced down the dragon's throat, "bursting it asunder." Exit one dragon. Interestingly, the great German explorer and animal collector, Hans Schomburgk, found a glazed brick in Central Africa—almost identical to the one discovered by Koldeway in Iraq. Could the ancient Babylonians have obtained their glazed bricks—and indeed, their dragon—from Central Africa, a place they were known to have explored?

Remarkable Artifacts

Today, on display at the British Museum in London, one can see a Mesopotamian cylinder seal dated at around 3300 BC. The animal on the right is an artist's conception of an *Apatosaurus* as it would have appeared in life. There are many striking similarities between these two depictions. The general morphology on the Egyptian art clearly fits a sauropod far better than any other type of animal. The biggest difference is at the head. Cartilage forming the shape of a head frill of some kind may be stylized or accurate, since there is no way to know from the fossilized skeletons that we have today. How did the artist fashion the size, shape, and musculature of a sauropod dinosaur so convincingly?

Another remarkable image can be found displayed on two slate palettes from Hierakonpolis showing the triumph of the early Egyptian ruler, King Nar-mer, depicting long-necked dinosaur-like animals along with numerous clear representations of known animals. The only logical conclusion that one can reach is that these images represent animals that we know today only from the fossil record. Yet the ancient artists who fashioned these images did not attempt to portray reconstructed fossil bones. These were clearly living animals—and animals that any modern child would instantly recognize as dinosaurs. Incidentally, the ancient Egyptians knew the behemoth by the name of *p'ih-mw*, which inhabited the Orontes River in Syria until after the time of Joseph (1900 B.C. – 1600 B.C.), and the lower Nile until the around the 12th century A.D.

Is there any further evidence that such a creature existed in Africa? Indeed there is! In the 1960s, a leading jewel designer, Emanuel Staub, was commissioned by the University of Pennsylvania to produce replicas of a series of small gold weights obtained in Ghana. So well crafted were they that the animals that they depicted could be instantly identified by zoologists—all but one, that is, which could not be satisfactorily reconciled with any known animal, until Staub saw it. Originally photographed resting on its hind legs (as if bipedal), this enigmatic Ashanti gold figurine was difficult to identify. Once properly positioned on all four legs that were of equal length, Staub noted that the mysterious artifact bears a striking resemblance to a dinosaur. The bulbous body, four strong legs, the thin neck and lizard- or snake-like head did not correspond with any known living animal in Africa today, or even any known creature that may have become extinct within the past 200 years during the era of European colonialism. Perhaps this figurine was an attempt to model a dinosaur that inhabited the remote regions of equatorial Africa.

Clearly, Africa and parts of the Middle East were home to extraordinary animals, including gigantic herbivores like the behemoth, which inhabited the swamps of the White Nile and possibly other locations in the Middle East, before either becoming extinct in those regions or withdrawing into more remote areas in central and southern Africa. This could possibly have been due to increased river traffic, lack of sufficient food supply, or being killed or driven away by humans. Whatever the case, reports of such monstrous creatures in the Middle East and North Africa eventually stopped and nothing more was heard of them until the colonization of Africa began in the 17th century by European powers.

COURTESY DR. ROY P. MACKAL / *A Living Dinosaur?*

Tanganyika Cave Art.

> In Africa, the past has hardly stopped breathing.
>
> —Trader Horn

2

THE DAWN OF ADVENTURE

During the 17th century, Belgium, France, Germany, Great Britain, and Portugal began in earnest to explore the African continent, and carved out huge chunks of territory for themselves. This in turn attracted adventurers, big game hunters, missionaries, traders, mineral prospectors, and settlers in enormous numbers. The Church of Rome sought to establish its presence in Equatorial Africa with zeal, and dispatched adventurous Jesuit priests to convert the heathen tribes that lived in the remote recesses of that mysterious continent. The Jesuits were keen students of nature, and when they were not seeking out converts to the faith, they would spend their time exploring the surrounding jungles, rivers, and lakes. Stories soon began to emerge from the natives concerning all manner of strange and wonderful creatures that inhabited the dense forests, uncharted rivers, forbidding swamps, and remote lakes.

In 1776, Abbé Lievain Bonaventure Proyart wrote in the *History of Loango, Kakonga, and other Kingdoms in Africa* about a group of French missionaries who had found the tracks of an enormous unknown animal while passing through the virgin equatorial forests. Pinkerton's translation, published in 1814, reads:

> "It must be monstrous, the prints of its claws are seen on the earth, and formed an impression on it of about three feet in circumference. In observing the posture and disposition of the footsteps, they concluded that it did not run in this part of its way, and that it carried its claws at the distance of seven or eight feet one from the other."

What could such a creature have been? The largest of African elephants have a similar stride but elephants do not possess clawed toes. This is one of the earliest written accounts alluding to the presence of large dinosaur-like monsters that began to emanate from most of French Equatorial Africa, particularly areas that possessed

dense forests, deep broad rivers, seasonally inundated swamps, and mysterious, remote lakes.

Ninety-four years later a young Englishman by the name of Alfred Aloysius Smith, later known as Trader Horn, arrived on the west coast of Africa to begin a career as a trader on the Ogowe (or Ogooué) River with the firm Hatton & Cooksons. Six decades later, in his old age, Smith scratched out a living by selling pots and pans door-to-door in South Africa. It was during this final career that Smith met South African novelist Ethelreda Lewis, who was so enthralled by the old man's recollections concerning his life as a river trader that she penned a book on Smith's life entitled *Trader Horn* (Simon & Schuster, 1927). Of interest to our search for living dinosaurs is Chapter 23, in which Smith recalls some details of a giant river monster greatly feared by the native peoples:

> "Aye, and behind the Cameroons there's things living we know nothing about. I could 'a' made books about many things. The *Jago-Nini* they say is still in the swamps and rivers. Giant diver it means. Comes out of the water and devours people. Old men'll tell you what their grandfathers saw, but they still believe it's there. Same as the *Amali* I've always taken it to be. I've seen the *amali's* footprints. About the size of a good frying pan in circumference and three claws instead o' five. There are some very big lakes behind the Cameroons. Used to be full of nice seal at one time. *Manga* they called it. But the *Jago-Nini's* wiped 'em almost out, the old natives say...."

Smith later continued:

> "That *Amali*, I told you I've seen a drawing of him in those Bushmen caves. I chiseled one out whole once and gave it to President Grant for a souvenir.... Aye, the little fellers that drew those creatures, and manacled slaves and so on I told you about, were not ordinary savages. They sure were paleolithic men from the North. They sure were remembering things, on those walls.... Nice little fellers, round about four feet and a little over."

The *amali* was clearly a formidable animal that remains unidentified for the time being. However, for the "seals" Smith mentioned inhabiting remote lakes, these were later identified by Dr. Roy P. Mackal, a biologist from the University of Chicago, as sea cows (*Trichechus senegalensis*), which were quite widespread in west African rivers in the 19th century. Dr. Mackal later conducted two groundbreaking expeditions to the Congo in search of our suspected dinosaur.

Also, the "paleolithic" men referred to in the passage were undoubtedly a pygmy tribe, probably the Aka people, which migrated from the densely forested region of the Congo basin to what is now Gabon. We shall meet these amazing people later in our travels through Africa.

Apart from the "narratives of the natives" as collected by Smith, what other sources of information can we refer to for reliability in these matters? Carl Hagenbeck (1844-1913), director of the Hamburg Zoo, and one of the greatest animal collectors of all time, was convinced that a dinosaur—or something akin to one—was still living in Africa. In his book *Beasts and Men*, published in 1909, he wrote:

> "Some years ago I received reports from two quite distinct sources of the existence of an immense and wholly unknown animal, said to inhabit the interior of Rhodesia. Almost identical stories reached me, firstly, through one my own travelers, and, secondly, through and English gentleman who had been shooting big-game in Central Africa. The reports were thus quite independent of each other. The natives, it seemed, told both my informants that in the depths of the great swamps there dwelt a huge monster, half elephant, half dragon. This, however, is not the only evidence for the existence of the animal. It is now several decades ago since Menges, who is, of course, perfectly reliable, heard a precisely similar story from the Negroes; and still more remarkable, on the walls of certain caverns in Central Africa there are to be found actual drawings of this strange creature. From what I have heard of the animal, it seems to me that it can only be some kind of dinosaur, seemingly akin to the brontosaurus. As the stories come from so many different sources, and all tend to substantiate each other, I am almost convinced that some such reptile must still be in existence. At great expense, therefore, I sent out an expedition to find the monster, but unfortunately they were compelled to return home without having proved anything, either one way or the other. In the part of Africa where the animal is said to exist, there are enormous swamps, hundreds of square miles in extent, and my travelers were laid with very severe attacks of fever."

Unfortunately, Hagenbeck died in 1913 without solving the mystery of the African "dragon" that had intrigued him for so long. But others would continue on the trail, determined to solve the mystery.

Carl Hagenbeck
(1844-1913)

Hans Schomburgk
(1880-1967)

Hans Schomburgk, the great German explorer and animal collector who worked in the employ of Carl Hagenbeck, traveled through a considerable amount of the forests of Equatorial Africa and wrote several books of his own including *Wild und Wilde im Herzen Afrikas*, published in 1910. Schomburgk was known as a meticulous recorder of the facts and did not take to fanciful stories easily. However, his experience as an explorer and animal collector in Africa led him to believe that there were many unknown species that remained undiscovered, particularly in and around the Congo basin.

Roy Mackal translated the following information relating to the mysterious water monster that allegedly inhabited Lake Bangweulu in Northern Rhodesia (now Zambia):

> "It is peculiar, that in Lake Bangweulu there appear to be no hippopotamuses. On the western side of the lake this could perhaps be explained, in that the banks are sandy and the hippopotamuses find the area unsuitable, although they are abundant in other reedless rivers. However, along the eastern and southern sides large swamps cover the area, which are certainly an ideal habitat for the hippopotamus, yet one rarely, if ever sees them there. It is certain that they have not been hunted to extinction. The natives say that an animal dwells in the lake, smaller than a hippopotamus, but which feeds on them. It must be completely amphibious, because according to the statements of the natives, it never comes on land and its footprints are never seen. I am sorry to say I regarded this story as a fairy tale and obtained very little exact information. Later I discussed the matter with Carl Hagenbeck and am now convinced that it must be some kind of saurian. Particularly since Hagenbeck received reports from independent sources which are in agreement with my information and with the reports of the natives in the region."

Stories of living dinosaurs in Africa began to capture the imagination of the public so much so that in November 1910, the magazine supplement of the *New York Times* asked in glaring headlines, "Is a Brontosaurus Haunting Africa's Wilds?" The front page revealed a somewhat inaccurate artist's representation of a sauropod dinosaur complete with two rhino-like horns on its elongated snout. Early black and white silent movies with the "stop-motion" models began to portray dinosaurs battling with each other and with early humans—in particular, primitive, grunting cave men.

Also in the early 20th century, King Lozi Lewanika of Barotseland (now part of northern Zambia) took a keen interest in the stories of a gigantic aquatic monster that lived in the marshes near his town. In order to see this very strange creature for himself, the king gave strict orders to his warriors that he be informed as soon as the beast made an appearance. A year later, three men rushed into Lewanika's courthouse to report that the monster had reappeared. The witnesses described an animal that was as big as an elephant, with strong legs like a lizard and a long slender neck ending in a small head like a snake. Unfortunately, the monster spotted its human observers and slid down the bank on its belly and disappeared into the marsh. Undeterred at such a missed opportunity to observe the monster for himself, the king rode to the area and observed the great breech made by the animal as it flattened down the reeds to make a path in the mud to and from the water. Later he made a report of his findings to the British resident, Colonel William Harding, commander of the Barotseland Native Police, that the breech made by the monster was "as large as a full-sized wagon would make with the wheels removed."

Shortly after the king reported his monster, the big game hunter Captain William Hichens made enquiries among the Barotse people in Northern Rhodesia about the monster, and learned that it was called *isiququmadevu*. Further investigation by Hichens uncovered reports of other lake dwelling monsters such as the *mbilintu*, said to inhabited Lake Bangweulu—now in modern-day Zambia, and Lake Mweru, which borders Zambia and the Democratic Republic of the Congo. Lake Tanganyika in Tanzania and the vast, unexplored swamps of the Congo were also reported to be home to a fearsome horned beast known locally as the *chipekwe*.

A careful examination of these reports clearly indicates that they speak of at least two different animals. The long necked, dinosaur-like *mokele-mbembe* clearly differs from the fearsome *chipekwe*, a hippo-sized animal that attacks and disembowels elephants with a single huge ivory horn on its nose. Like Von Stein and Schomburk before him, Hichens interviewed a number of natives from differing religious, ethnic, and cultural backgrounds and found that they all described the same animals but with different names in each individual dialect. Clearly, these were living animals that deserved further investigation.

This and similar reports undoubtedly inspired the novel *The Lost World*, authored by Sir Arthur Conan Doyle and published in 1912. The book chronicles the adventures of four explorers led by a certain zoologist named Professor Challenger, whose claim to have found a colony of living dinosaurs and a race of ape-men on a remote plateau mountain in South America is greeted with incredulity by the scientific establishment of the day. Undeterred by the unscholarly rejection of his astonishing claims, Challenger organizes another expedition, which included the aristocratic explorer,

Lord John Roxton. After a hair-raising adventure, the explorers return to London with a captive pterodactyl, which Challenger later gleefully released in a room filled with his skeptical colleagues.

By the time Conan Doyle had published his famous novel, the Darwinian concept of long geologic ages had begun to establish its iron grip in the minds of the masses as the young field of paleontology began to uncover fossilized dinosaur bones for display in museums around the world. In spite of the secularization of science, reports still reached the Western World that "monstrous creatures" were still very much alive in Africa.

But how far-fetched is Conan Doyle's story? Did he base his alleged work of fiction on actual events? As adventurers, explorers, missionaries and colonial administrators began to penetrate deeper into the dark recesses of Africa, more tales of terrifying monsters and cannibal pygmies began to stock the fertile imaginations of armchair explorers around the world. Would some fortunate and daring soul eventually find the primeval monsters that stirred in the malaria-ridden swamps and uncharted rivers?

> There are lost worlds everywhere.
> —Bernard Heuvelmans

3

THE DRAGON HUNTERS

In 1913, the German government decided to survey Cameroon (Germany's colony at the time), and chose Captain Freiherr von Stein zu Lausnitz to lead the expedition. Von Stein included the following fascinating report on a creature "very much feared by the Negroes of certain parts of the territory of the Congo, the lower Ubangui, the Sangha, and the Ikelemba Rivers.... They call the animal *Mokele-mbembe*." Von Stein's expedition was cut short due to the commencement of World War I, but part of his report was later translated into English by German rocket scientist Herr Willy Ley. The relevant passage reads:

> "The animal is said to be of a brownish gray color ... its size approximating that of an elephant. It is said to have a long and very flexible neck and only one tooth, but a very long one; some say it is a horn. Some spoke of a long muscular tail like that of an alligator. Canoes coming near it are said to be doomed; the animals are said to attack the vessels at once and to kill the crews but without eating the bodies. The creature is said to live in the caves that have been washed out by the river in the clay of its shores at sharp beds. It is said to climb the shore even in daytime in search of its food; its diet is said to be entirely vegetable. This feature disagrees with a possible explanation as a myth. The preferred plant was shown to me, it is a kind of liana with large white blossoms, with a milky sap and apple-like fruits. At the Ssombo River I was shown a path said to have been made by this animal in order to get at its food. The path was fresh and there were plants of the described type near by."

Von Stein took a skeptical, levelheaded approach like Hagenbeck before him. One can only speculate what important discovery Von Stein could have made if

World War I had not cut short his expedition. After all, he was in the target area and had examined fresh evidence of the mystery monster's determination to reach its food supply, which grew there in abundance. Had time not been against Von Stein, an extended stay, coupled with careful river exploration in the immediate vicinity of the mystery animal's food supply, might have heralded results that could have delivered the creature into his hands and solved the mystery for good.

Stories of possible living dinosaurs in Africa inevitably attracted a few clever hoaxers, intent on pulling the wool over the eyes of the public, it seemed, for no other reason than to enjoy a good belly laugh. On November 17, 1919, the following report was published in the *Times*, under the headline, "A Tale From Africa, Semper aliquid novi." The central news Port Elizabeth correspondent sent the following:

> "The head of a local museum has received information from a M. Lepage, who was in charge of railway construction in the Belgian Congo, of an exciting adventure last month. While Lepage was hunting one day in October he came upon an extraordinary monster, which charged at him. Lepage fires but was forced to flee, with the monster in chase. The animal before long gave up the chase and Lepage was able to examine it through binoculars. The animal, he says, was about 24 ft. in length with a long pointed snout adorned with tusks like horns and a short horn above the nostrils. The front feet were like those of a horse and the hind hoofs were cloven. There was a scaly hump on the monster's shoulders. The animal later charged through the native village of Fungurume, destroying the huts and killing some native dwellers. A hunt was at once organized but the government has forbidden the molestation of the animal, on the grounds that it is probably a relic of antiquity. There is a wild trackless region in the neighborhood which contains many swamps and marshes, where, says the head of the museum, it is possible that a few primeval monsters may survive."

The description of a swamp dwelling, hippo-killing monster sporting a single large horn on its nose reminds one of the fearsome *chipekwe*, which reportedly inhabited the vast swamps of the Belgian Congo at that time. However, the "scaly hump," tusk-like horns, a short horn above the nostrils, cloven hooves and a length of 24 feet make it sound quite ridiculous. It soon became apparent that this description did not fit any known animal, living or extinct, and the report was beginning to smell like a hoax anyway. Unfortunately, the jackelopian description of the mystery animal was given some unintentional credibility as further reports of alleged living dinosaurs appeared with the pages of one of America's most prestigious newspapers.

Four days later, on December 13, 1919, the *New York Times* published the following article, headlined: "Hunter Says He Saw Prehistoric Monster. Belgian Returns from Congo with a Story of Tracking Down a Brontosaurus."

> "Details of the alleged discovery of a survivor of the prehistoric Brontosaurus are given in a delayed Reuter message from Bulawayo by the Belgian prospector and big game hunter, M. Gapelle, who has returned there from the interior of the Congo States. He followed up strange footprints for twelve miles and at length, he says, sighted a beast of the rhinoceros order with large scales reaching far down its body. The animal he says, had a very thick kangaroo-like tail, a horn on its snout and a hump on its back. Gapelle fired some shots at the beast, which threw up its head and disappeared into the swamp. It was pointed out that the American Smithsonian expedition was in search of the monster referred to above when the members of the party met with a railway accident in which several persons were killed. Discussing the Smithsonian expedition yesterday, Dr. S. F. Harmer, director of the zoological section of The British Museum of Natural History, while keeping an open mind said: 'I should doubt strongly the survival of any of the race of Dinosaurs, the enormous creatures belong to so remote an age. In the forest and swamp of untrodden portions of Africa there are probably remarkable specimens of unknown beasts, but the likelihood is they are mammals of a much later type. The discovery in modern times of remains of extinct animals has all been of creatures comparatively recently extinct.'"

However, in spite of the excitement these and other similar reports generated among the general public, the bubble soon burst when the following letter was published in the *Times* on February 23, 1920:

> "Sir, I am authorized to contradict the statement that the members of the Smithsonian African Expedition who proceeded to this territory came here to hunt brontosaurus. There is no foundation for this statement. I may also state that the report of brontosaurus arose from a piece of practical joking in the first instance, and, as regards

the prospector 'Gapelle,' this gentleman does not exist except in the imagination of a second practical joker, who ingeniously coined the name from that of Mr. L. Le Page.

"Yours Faithfully,
 Wentworth D. Gray
   Acting Representative of the Smithsonian
   African Expedition in the Katanga
   Elizabethville, Jan. 21."

Hoaxes of this kind did not deter well meaning but somewhat naïve adventurers like Captain Leicester Stevens, a decorated veteran of the Great War (1914-1918). Stevens was a tall, handsome Englishman somewhere in his 30s and a romantic adventurer to boot. On the morning of December 23, 1919, Captain Stevens left London's Waterloo Station, bound for a waiting Ship at Southampton. "I am leaving for Cape Town on Christmas Eve," he told newspaper reporters. "From Cape Town I shall go 1,700 miles north to Kafue, where my expedition will be organized." The dashing captain had as his only companion a mongrel half-wolf called Laddie, which had delivered messages to allied soldiers while under artillery fire. Laddie's job was to track down the beast, whereupon his master would kill the monster with a well-aimed shot from his Mannlicher rifle. The alleged brontosaurus, the press was told, possessed a vital spot. "Where that vital spot is" stated Stevens, with a twinkle in his deep blue eyes, "is one of my secrets." One can only speculate just where the "vital spot" on a dinosaur bigger than five elephants might be, particularly as the Mannlicher rifle that Captain Stevens possessed was barely able to kill a bear.

Nothing much was heard of Captain Stevens after his dramatic departure from the green fields of England. However, the more sensational tabloids of the day published a brief account of how Stevens allegedly spotted a gigantic brontosaurus "bigger than ten elephants" plunging through the swamps. I personally believe that our latter day Saint George failed to find his dragon and returned home far more discreetly than when he departed.

But were the rumors of living dinosaurs—or at least gigantic animals that looked like them—mere native stories supported by the occasional clever hoax?

By the 1920s improved travel conditions in Africa by air, rail and river continued to develop, bringing more reports of possible living dinosaurs to the ears of the colonial powers. In 1923, a long-necked "prehistoric" beast was observed by officers onboard the Victoria steamer, the *Nyanza*. Not long after this exciting event, the monster was spotted again, this time by Sir Clement Hill, who was traveling from Kisumu to Entebbe onboard a steam-powered launch.

During the journey, Sir Clement was enjoying a pleasant view of Homa Mountain when his attention was arrested by the sight of a terrifying monster that suddenly appeared from the depths of the lake and attempted to seize a native sitting at the bow of the vessel. The would-be victim managed to fend off the strange creature, which disappeared back into the depths of the lake.

Here were solid, eyewitness accounts of a strange or "monstrous" animal that had been observed by a handful of westerners—the officers of a large steamer and a knight of the realm—individuals of integrity who had made important observations that gave considerable credibility to the native reports that an aquatic monster of some kind undoubtedly inhabited Lake Victoria. If such a monster did indeed kill human beings as well as hippos from time to time, then the natives had good reason to feel vulnerable as they fished in the lake every day and traveled about on it in flimsy canoes that offered little protection from the beast.

A "wild country that speaks of nightmare and violent death..."

—Unknown

4
TALES FROM THE HEART OF DARKNESS

In 1930, Lucien Blancou, the French chief game warden in the Likouala region of French Equatorial Africa (which later became the Republic of the Congo), collected numerous reports of the strange animals that allegedly haunted the forests, river, and swamps of the territory in which he worked.

Among the frequent reports that caught his interest was of a gigantic "snake" that killed hippos without leaving any wounds or marks on them. The "snake" was also said to browse on the leaves that grew on the branches of the trees along the river without leaving the water. Leaves would be a very odd diet for a snake, and for any snake to elevate itself to the high branches would be quite a feat unless it was of gigantic size and resting on the bottom of the river.

Chief Mapouka, head of the Yetomane division, was a great hunter and thoroughly at home in the bush. Regarding the mystery animal, the chief told Blancou that one such creature ventured onto dry land one night in 1928 and crushed a field of manioc belonging to the chief, leaving huge tracks between three and five feet wide. A young Banziri secretary also informed Blancou that his people referred to the same animal as *songo*. The same monster was said to have killed a hippopotamus at the same time in the river Brouchouchou near the chief's village, but in spite of the great fear the local populace had for the "water devil" they had no qualms about eating its prey!

The people of the Baya tribe knew the same animal, which they called *diba*, that inhabited the river in the area of Bozoum in 1934. During his enquiries into the animal, Blancou was given more precise information from an old villager called Moussa, who was brought to Blancou because he was a first hand eye-witness, having seen the animal for himself. Moussa said that his people knew the animal well and that it was called *bagidui*. Monsieur Blancou recalled Moussa's encounter with the fabled monster:

> "When he was about 14 year old and the whites had not yet come (about 1890 I suppose), Moussa was out laying fish traps with his father in the Kibi stream, which runs into a tributary of the Ouaka called the Gounda in what is now the Bakala district. It was one o'clock in the afternoon in the middle of the rainy Season. Suddenly Moussa saw the *bagidui* eating the large leaves of a *roro*, a tree, which grows in forest-galleries. Its head was flat and a bit larger than a python's (Moussa spread his hands and put them together to show me the size). Its neck was as thick as a man's thigh and about 25 feet long, much longer than a giraffe's; it had no hair but was as smooth as a snake with similar but larger markings. The underneath of its neck was lighter also like a snakes. Moussa did not see the body. His father told him to follow him and run away... Finally, he said that the *badigui* does not frequent places where you find hippopotami, for it kills them."

There is no doubt old Maoussa exaggerated the neck length at 25 feet. A fully-grown male giraffe is up to 18 feet in height, but the *badigui* would most likely have a neck length of 10 or 12 feet, which would be more realistic for an animal with a body the size ranging from a hippo to a forest elephant (though the fossil record does show more bizarre proportions in creatures such as *Mamenchisaurus* and *Tanystropheus*).

In 1932, Scottish-born American zoologist Ivan T. Sanderson and American naturalist Gerald Durrell were engaged in collecting various specimens of flora and fauna as part of the Percy Sladen Expedition. Their travels took them through a large part of west and central Africa and eventually into Cameroon. As the explorers slowly made their way into the Manyu River with their young native guides, Besun Onum Edet and Bassi Aga, they witnessed an extraordinary sight, the details of which were later published in Sanderson's book, *More Things*:

> "On returning downstream to the Mamfe pool from a day of very hard paddling upstream on a collecting trip, we just glided a long, paddling only now and then to maintain way. Sundown was approaching as we entered the gorge. Gerald was in the lead canoe with Bassi; I followed about a hundred feet behind with Bensun. There were deepening shadows in the gorge and all along its towering vertical walls were the arched tops of huge caves. We had previously penetrated these at the pool end of the gorge to collect a certain kind of very

rare frog, but we had never before passed these huge ones farther upstream. When we were in the middle of the mile-and-a-half-long winding gorge, the most terrible noise I have heard, short of an oncoming earthquake or the explosion of an aerial-torpedo at close range, suddenly burst from one of the big caves to my right. Ben, who was sitting up-front in our little canoe with a 'moving' paddle, immediately dropped backward into the canoe. Bassi in the lead canoe did likewise, but Gerald tried to about face in the strong swirling current, putting himself broadside to the current. I started to paddle like mad but was swept close to the entrance of the cave from which the noise had come. Thus, both Gerald and I were opposite its mouth; just then came another gargantuan gurgling roar and something enormous rose out of the water, turned it to sherry-colored foam, and then, again roaring, plunged below. This 'thing' was shiny black and the head of something shaped like a seal but flattened from above to below. It was the size of a hippopotamus, this head I mean.

"We exited the gorge at a speed that would have done credit to the Harvard Eight and it was not until we entered the pool that Bassi and Ben came-to. What we wanted to know, what was this monster? Neither could enlighten us, as they were not river people. However, both finally yelled 'M'koo-m'bemboo,' and grabbed their paddles. When we reached the little beach at the far end of the pool where we kept our canoes, we were met by the rest of our gang, some twenty-strong and all local men. They were very shaken and solicitous of our safety. All the river people among them confirmed bassi and Ben's diagnosis. These animals live there all the time, they told us, and that is why there were no crocodiles or hippos in the Mainyu. (*sic*) (There were hundreds in the pool, the other river and the Cross River). But, they went on, 'M'koo' does not eat flesh but only the big liana fruits and the juicy herbage by the river. Later we moved across the river permanently and camped nearby. We found huge pathways through the herbage from the river and masses of the great, tough, green 'footballs' all smashed up, and some with pieces, a foot wide, bitten out of them Just as we bite a piece out of an apple."

In 1971, Sanderson corresponded with James Powell, a crocodile expert from Lubbock, Texas, who had studied the rainforest crocodile population in Gabon. In his letter to Powell, Sanderson stated with regard to the monster he saw, "its head was bigger than a whole hippo and the tracks were sauropod."

The problem with this statement is that sauropods are known to have small heads, and any creature that possesses a hippo-sized head cannot be a sauropod. Sanderson and Russell could have actually seen the back of the animal break the surface, and mistaken it for the head. Also, later research found that the name *mokele-mbembe* is sometimes used in reference to any strange or unknown animal rather than our long-necked mystery animal. In this case, it is quite plausible that if Sanderson and Russell really did see an actual seal-like hippo-sized head of some monster, the confusion with the *mokele-mbembe* proper could be explained. What is most intriguing about this monster is that the Ayang people that lived in that area said it was an herbivore that drove away hippos and even crocodiles, all factors that are consistent with *mokele-mbembe* proper. It seems likely, given the other common characteristics, that the back of the animal was seen, rather than the head.

In 1938, six years after the Percy Sladen Expedition, Dr. Leo Von Boxberger, a former colonial magistrate with considerable African experience, picked up stories about a much feared water monster and later wrote:

> "My own contribution to the subject is unfortunately very small. At the mouth of the Mbam in Sanaga in central Cameroons and on the Ntem in southern Cameroons, I collected a variety of data from the natives about the mysterious water beast, but, alas, all my notes and the local description of the animal were lost in Spanish Guinea when the Pangwe tribe attacked the caravan carrying my few belongings. All that I can report is the name *mbokalemuembe* given to the animal in Southern Cameroons... The belief in a gigantic water animal, described as a reptile with a long thin neck, exists among the Natives throughout the southern Cameroons, wherever they form part of the Congo basin and also to the west of this area, doubtless wherever the great rivers are broad and deep and are flanked by virgin forest."

The loss of Boxberg's notes was indeed a blow to those who eagerly sought out clues to the mystery animal's identity, but nevertheless his brief but interesting report confirmed the fact that a mystery water beast, approximating the size of an elephant was still very much alive and well in remote rivers of southern Cameroon and the northern Congo, a difficult, hostile, and remote part of Africa where the human population is very small, and a perfect hiding place for a giant water monster!

Given these reports, it is obvious that some strange, long-necked, aquatic animal and other strange creatures do indeed inhabit the rivers, swamps and lakes of west and Central Africa. The different ethnic, cultural, and tribal groups may all refer to the animal by different names such as *amali*, *badifui*, *dibi*, *embulu-mbembe*, *isiququmadevu*, *jago-nini*, *songo*, *m'koo-m'bemboo*, and *mokele-mbembe*, but the general description of the animal, including its semi-aquatic habitat, bulbous body, long thin neck, small snake-like head (sometimes adorned with a frill like a rooster), long flexible tail, and four powerful legs really does make it sound like a dinosaur.

Very little else was heard of our mystery beast between 1939 to 1945, as World War II raged around the globe. The few isolated reports that did reach the French colonial authorities were largely ignored due to the impact that the war was having on the world at large. But the monsters stirring in the dark, forbidding swamps would not remain forgotten for very long. In 1948, public interest again was kindled in the subject when *The Saturday Evening Post* ran an article written by Ivan T. Sanderson entitled "There Could be Dinosaurs," in which Sanderson wrote of his belief that dinosaurs might still be alive in equatorial Africa.

The article also caught the interest of Bernard Heuvelmans, a young French zoologist and jazz musician who in later years would become widely known as the "father" of cryptozoology. Heuvelmans, who later earned a doctorate in his field, would go on to author some of the most influential books ever written on the subject of mysterious and unknown animals, based on a lifetime of research.

That same year, another fortuitous occurrence involved multiple witnesses, but with only one being named for the record. In 1948, Air Security Officer A. S. Arrey was keeping some British soldiers company as they swam in Barombi Mbo, a crater lake near Kumba in northern Cameroon. Suddenly the swimmers stopped their activities as the water in the center of the lake began to stir, as if being disturbed from below. After everyone had hastily vacated the water, two strange long-necked creatures broke the surface. The larger animal had a neck estimated to be 12-15 feet long, ending in a small slender head that sported a spike of some kind just behind the head. A second, slightly smaller animal also emerged from the lake, but did not sport an appendage or adornment of any kind. The locals later claimed that the smaller of the two animals was a female. Arrey recalled that a few of his British companions deserted the scene, but some remained and kept the animals under observation. Although the animals closely resembled the *mokele-mbembe*, they possessed scales similar to a reptile. Later reports from southern Cameroon supported the idea that a larger, mature *mokele-mbembe* possesses a toughened skin much like a caiman.

In spite of the great fear exhibited by most of the native peoples who are familiar with our water monster and its alleged invincibility, there have been a few, isolated cases where they have been killed by village hunters. One such report was sent to Dr. Roy Mackal from Jorgen Birket-Smith, a Danish national who had spent the winter of 1949-1950 in the (then) French Cameroons, in a place called the *Case du Nyong*, located on the River Nyong and run by the *Service des eaux et forts*. Birket-Smith recalled the following details during an interview he conducted with an elderly hunter called Jerome and a second native referred to only as a *Guard de foret*:

> "They both knew a big animal living in the river (Sanaga), which they called 'Nwe.' I drew a hippo. Yes, they knew that one, but that lived in Nyong by Mbalmayo, which was quite correct. I then drew a crocodile. Yes, they were in the Nyong just by the house and not very big. I explained, they could be many meters long, and they agreed, for in the Sanaga they had seen such big ones. Then they explained that the Nwe had a long neck and I drew a giraffe. That one they also knew, but that one never swam in the river, and one should far up north to find it. Since I knew many of the older records of a 'Brontosaurus,' that should live in Africa, I drew one. They immediately both agreed that this was like a Nwe. Having arrived that far, I began to pump them for all they knew about the Nwe. It was very rare, or at least very rarely seen. However, the guard remembered, when he was a boy—which must have been sometime in the twenties that they had caught one in his village. It size was between that of a hippo and an elephant. The whole village had eaten from it in a full week. (Must have had a nice odour after some days!)—but that is of no account to most Africans. ... Of the habits of the Nwe they could not tell much, only one interesting thing, it ate from the trees, by which they meant the trees in the gallery forest leaning over the water. ... It hardly ever came on dry land."

Collectively, these reports confirm that these animals were well-known to the tribespeople who lived throughout the Congo, Cameroon, Ubangui Shari, Sudanese marshes, and Zambia. These areas typically had a thin human population, coupled with rivers, streams, lakes, and swamps which remained tranquil for the most part, with little or no human encroachment into such areas. The only exceptions were those who fished, hunted, and used the river for traveling from one place to another.

In spite of the scarcity of western observers of these extraordinary animals, fortune favored at least three more eyewitnesses, whose accounts would prove to be extremely important pieces of the puzzle in helping to identify our mystery monster.

In May 1954, Englishman Alan Brignall was working in Kiture, Northern Rhodesia (now Zambia), when he and his colleagues decided to take a fishing trip to Lake Bangweulu. That hot, sunny afternoon, Brignall happened to glance out on the lake towards two small islands about 25 yards distant that were surrounded by reeds. The Englishman was thunderstruck by the sight of what he first thought was a huge snake that had risen out of the water near the shore of one of the islands. But Brignall quickly realized that animal was quite unlike any familiar snake. On observing the monster, Brignall noticed that it possessed a small snake-like head at the end of a long smooth, gray colored neck about a foot wide and raised some 4½ feet out of the water. The facial features of the animal possessed a prominent brow, blunt nose and clearly visible jaw line. Fascinated by the appearance of this amazing animal, Brignall watched as the head of the creature swiveled from side to side as if looking for something. Behind the neck was a visible hump, which Brignall thought may have been a small portion of a much larger body underwater.

Recalling his story decades later, Brignall believes that he alerted the animal either as he tried to signal to a friend who was fishing nearby or as he attempted to reach for a camera. Either way, his movement caused the animal to silently sink into deeper water where it remained hidden. If there was one thing Alan Brignall was certain of, it was that the amazing creature he observed was not a snake, a crocodile, a hippo, or any other known animal that lived in that region.

Going by Mr. Brignall's careful description of the animal he observed, it is clear that what he saw was a *badigui* of the long-necked variety, possibly a close relative of the *mokele-mbembe* of the Congo, and still present in the lake until fairly recent times. Lake Bangweulu is relatively shallow, but is surrounded by vast swamps that would allow any large animal to move to and from the lake undetected.

The second eyewitness account involved Mr. George Shepherd (not his real name), who was working in the Belgian Congo near Stanleyville (now Kisangani). Early one evening while relaxing on the verandah of his quarters overlooking the mighty Congo River, Shepherd's attention was drawn to some movement in the tree tops along the edge of the river. Thinking that perhaps a troupe of monkeys was moving through the trees, the Englishman was astonished to see the long thin neck of some strange creature moving among the treetops, browsing on the vegetation. Although the animal was about 200 yards distant, Shepherd could clearly see that this was no animal with which he was familiar, describing it as "a long thin neck with a set of jaws on the end." The animal also possessed a large, bulky body that broke the surface occasionally as it moved along. Its color was "gray all over." I am indebted to Mr. Shepherd's family passed on to this report to me on guarantee of his anonymity.

The third important observation of a possible *mokele-mbembe* was made in 1961, reported in the form of a letter, and mailed to Dr. Roy Mackal at the University of Chicago, from J. M. Lefebvre, who was a resident in Pretoria, South Africa. The letter, dated July 14, 1981, is reproduced here with Dr. Mackal's kind permission:

"Dear Sir,

"Will you please forgive my bad english, my home language is French, and I still have some communication problems—*The Pretoria News* have communicate in your address to me, asking me to contact you. In the first place, I am referring to 20 year old events, that will account for the lack of precision of certain details like exact location and so forth. The sighting took place in an area situated between 20° and 28° long. East and 0° and 4° lat. North. The country was rather marshy, pools of mud divided by narrow stretches of dry land. It was a vast clearing, approximately 1-mile diameter. The vegetation was tall elephant grass, reeds and papyrus. The surroundings of the clearing was the tropical rain forest; trees about 90 feet high. Time 13:00 h, Date July 1963. I first came across the spoor in a muddy spot. It was formed by, in the center a kind of depression or farrow between 3 and 6 feet wide, similar in shape to the one one would make by dragging a bag of coal in the sand. On each side of it were footprints, 2 to 3 feet wide by 3 to 4 feet long, egg shaped with the broadside towards the front. Imprinted deep into the mud was the mark of five ribs or fingers approximately 3 inches wide, starting from a common centre at the back of the print, diverging towards the front and sides. The space between the fingers was also imprinted with something softer, the ground was raising slightly. I would compare it to the diaphragm between the claws of a duck—the footprints were between 3 and 5 feet on each side of the belly mark and the marks of the rear feet were overlapping the ones of the front one. The tall shrubs which were on the spoor were crushed like with a bulldozer. At the sight, my trackers stopped abruptly and refused to let us go farther. Considering the size of the animal and the gauge of the guns, my friend and I decided to retreat—On our way back, the trackers showed us on the other side of the clearing the animal. However, at about 1 mile distance and with the haze it was not very clear—A huge greyish mass, towering well above the grass. I estimated from 20 to 40 feet high. A long flexible neck about the size of a tree trunk + 30 feet long. A very small head held at a right angle from the neck. I had a camera with me, and I did not take a photo—in the first place I never thought about it, in the second place because at this distance and with the haze, it never works. The friend who was hunting with me, Lem Bauer, had been killed by the Simbas, the local terrorists."

The animal observed by Lefebvre and Bauer must have been an extraordinary sight! The impressive size of the creature, coupled with the length and thickness of its neck, its height, the size of the tracks made in the ground, the enormous breech through an area of shrubbery and river grass, and the obvious fear of the native trackers over this discovery and subsequent observation, even from a distance of one mile, argues strongly in favor of a very large and mature specimen of *mokele-mbembe*. LefeBvre and Bauer were experienced hunters and obviously knew that the monster they tracked and later observed was quite different from any kind of familiar African animal. Furthermore, the area where the two hunters made their observation was close to the location where Jorgen Birket-Smith had learned about the dinosaur-like *nwe* some ten years before, where the local population was quite familiar with them.

These important observations, made by educated Europeans, not only add considerable weight to the reports by the local tribes that were familiar with such animals, but support the idea that the *badigui* and the *mokele-mbembe* were still present in lakes and rivers of the Central African Republic and the former Belgian Congo until fairly recent times, and may still inhabit these areas today. It would be interesting to see what a modern-day expedition to either of these locations would uncover. However, at least two more (known) sightings of *mokele-mbembe*—or at least animals that look very much like them—and one report involving multiple witnesses, would find their way out of Africa and into the pages of our ever-growing file on suspected living dinosaurs.

In 1955, Eugene P. Thomas, and his wife, Sandy, from Wisconsin, USA, arrived in the town of Impfondo in the northern Congo to begin their careers as long-term missionaries. The young couple faced incredible obstacles. Eugene was tasked with establishing the first Protestant church and Bible school in the country, which was heavily influenced by the Catholic Church and the then French colonial administration. Sandy patiently labored to build a clinic for the treatment of all manner of tropical diseases and maladies, including leprosy. The work was extremely challenging and for the first five years the young couple lived in a native house with dirt floors. As Pastor Thomas spent time getting to know the local Congolese residents in his efforts to establish a bush church, he began to hear stories about several strange

and unusual animals that inhabited the forest and swamps. However, the young missionary was not interested in exotic wildlife—his work was winning souls for the Lord. A few years later, though, two old pygmy hunters told him an extraordinary story that stuck in his mind.

It was around 1960, and a tribe of Bagombe pygmies spent their years hunting and fishing in and around Lake Tele, a shallow, oval body of water located in the heart of the Likouala Swamp. However, the tribe was growing weary of having its fishing activities disturbed by two or three *mokele-mbembes* that would move into the lake on a daily basis from the swamps via a *molibo*, or water channel. In spite of the fact that the pygmies were very afraid of the animals, they decided to stop the disruption in their fishing routine and erected a stake barrier across the entrance to the lake—then waited. One morning, two of the animals were observed trying to break through the barrier in order to enter the lake. The pygmies, armed with their 10-foot-long elephant-killing spears, immediately attacked one monster and succeeded in spearing it to death. The tribe then butchered the animal for its meat, which, apparently, was as bountiful as an elephant. However, after killing and butchering, those who ate some of the meat of the animal died shortly afterwards. This may have been because the meat of the animal contained toxins that were fatal to humans, or perhaps food poisoning. (John Kirk suggested the possibility of Salmonella, which is not uncommon in cases dealing with reptiles.) Or it may have been coincidence, as the simple fact is that forest dwelling pygmies often do not live much beyond 35 years of age. As the oldest and most experienced hunters of the tribe would have been chiefly responsible for leading the attack in the slaying of the creature, they would also have been the first to consume the meat due to their seniority in the tribe.

In any case, the short lifespan of the traditional forest pygmy would indeed have a bearing. Pygmy women give birth from the age of twelve, and their hunter-gather lifestyle, coupled with vulnerability to all kinds of dangers—diseases, infections, venomous snakes, and wild animals—would almost certainly contribute to the dangers of living in such an environment. The killing and eating of a suspected *mokele-mbembe* or *nwe* in the French Cameroons some forty years before by a different (Bantu) tribe did not herald any deaths or serious illnesses, as far as we know. Nevertheless, the Lake Tele incident quickly developed into a belief of sorts among the present-day inhabitants of the Likouala Region of the Republic of the Congo, that if one sees or encounters a *mokele-mbembe*, then speaks openly of their experience—great misfortune or even death will result.

Not only was the killing of a *mokele-mbembe* at Lake Tele an interesting story in itself, but Pastor Thomas later recalled how the two elderly pygmies described the killing in detail, and also mimicked the strange, high-pitched cry of the monster as it was being speared to death. Although all the reports on *mokele-mbembe* tell us that it is usually a silent animal, it is common for many animals to cry, roar, shriek, or bellow when under attack, especially by a group of expert spear-throwers intent on killing their prey with 10-foot long elephant spears. Considering the ferocity of the attack, the severity of its wounds, and rapid loss of blood, any animal, no matter how big, would have made some form of extreme vocalization, especially if it had never before encountered this kind of confrontation with animal or human, and experiencing such alarming pain.

Other stories would surface from time to time, conveyed to Pastor Thomas in hushed tones, lest the *N'dami* (elders) of the villages find out and punish the informants. Although the natives were very much afraid of *mokele-mbembes*, occasionally they would kill very young specimens if they were encountered alone in the river and without the presence of a protective adult. This sounds perfectly plausible, considering that the natives often kill sizeable crocodiles, hippos, and even elephants with nothing more than spears. If *mokele-mbembes* are real flesh-and-blood animals, then they can also be killed by a group of determined (or frightened) hunters.

Although there are very few reports of *mokele-mbembes* being killed or remains of the animals ever being found, another elderly native resident of Impfondo related a story to Pastor Thomas about how a dead *mokele-mbembe* was found partially submerged in a swamp near the town Epena, located on the Likuoala-aux-Herbes River. The incident happened around 1938, and cause of death was uncertain, but the animal could not have been dead for very long, as it was still intact and without any outward signs of decomposition. A couple of French engineers tied a chain around the tail of the animal and attempted to pull the beast out of the swamp pool with an old bulldozer, but part of the tail broke away from the rest of the body and the remains slowly sank into the depths of the swamp—lost forever.

Among the few Europeans to find secondary evidence of *mokele-mbembe* activity was Monsieur Langois, a French storekeeper who lived near Impfondo. Sometime in the early 1950s, Langois was on a fishing expedition to the Likouala-aux-Herbes River just south of Epena, when he came upon huge elephant-like footprints that led out of the water and along the riverbank, as if some gigantic animal had been foraging for food. He rushed back to his store, retrieved some plaster-of-Paris and managed to make two or three very detailed casts of the footprints before they were destroyed by the rains.

After acquiring the casts, Langois excitedly sent a telegram to the French colonial administration in Brazzaville stating that he had found fresh "dinosaur tracks" on the bank of the Likouala-aux-Herbes River and had made some detailed casts for them to see. The authorities, however, were less than impressed by the claims of a common storekeeper and accused Langois of being a drunkard! The French governor of the territory at the time, Paul Louis Gabriel Chauvet, was more concerned with maintaining order in what was then known as Middle Congo prior to eventually handing power over to an independent Congolese government without the country collapsing into civil war.

Unhappy about the attitude of the colonial government towards his discovery, Langois gave up trying to generate interest in his plaster casts. Fate dealt a cruel blow to Monsieur Langois when he decided to return to his native France after Congo-Brazzaville became an independent state in 1960. Moments before his plane landed in Paris, Langois suffered from a fatal heart attack. The man and his plaster casts were lost forever.

> There is always trouble
> when a white man is killed.
>
> —Chief Mokaka,
> Epena District

5
THE EXPEDITIONS 1979-1992

After 1960, very little else had been heard of *mokele-mbembe* or any other mystery animal in west and central Africa, apart from vague rumors from big game hunters and adventurers. The former Belgian colony of the Congo was in the grip of a civil war that claimed untold thousands of lives. Almost all of the African colonies under France and Great Britain had become independent states in their own right, with wars raging in Algeria, Angola, Congo, Nigeria (Biafra), and Mozambique. Unrest even rocked the tiny Indian Ocean island of Mauritius following its independence from Great Britain in 1965. Many newly independent African states were turning their backs on their former colonial masters to embrace extreme socialist policies, and the Rhodesian bush war was beginning to heat up. By 1970, there were so many wars in Africa that the western media could barely keep up with them. Expatriot Europeans and even many missionaries fled Africa in fear for their safety, but some, like the Thomas family in the Congo, stayed on, determined to continue with their work regardless of the circumstances or the many dangers to their own safety.

As Gene Thomas continued to train indigenous pastors at the Bible school he established, stories about *mokele-mbembe* kept surfacing as some of his students—especially those from the more remote villages located on the river system in the northern Congo—related some of their own experiences to him. One such report was volunteered by an old village chief who lived in the bush west of the town of Epena. The chief told Thomas that there was a large swamp pool about "a day's walk" from his village. There, he and his warriors would sometimes observe a "giant animal" that would emerge from the swamp and climb onto a small island, where it would beat its huge tail on the ground, making the earth literally shake. Intrigued by this report, the adventurous missionary, accompanied by his pygmy guides, was ready for a trek into the jungle to see this monster for himself. However, the old chief, afraid that Thomas would lose his life to the beast, resolutely refused to let him go.

THE MACKAL EXPEDITIONS

In 1976, James H. Powell, Jr., a herpetologist from Plainview, Texas, was studying crocodile populations in Gabon, West Africa, when he picked up a few scattered reports about a giant water dwelling monster called the *n'yamala*. After returning to the United States, Powell decided to attend a lecture in Amarillo, Texas, delivered by Dr. Roy Mackal on his many years of scientific research at Loch Ness. Mackal was also vice-president of the International Society of Cryptozoology, based in Tucson, Arizona, which made him by far the most qualified academic researcher on the subject of mysterious animals.

Powell approached Dr. Mackal at the end of the question-and-answer period and introduced him to the reports concerning the *n'yamala*. Mackal listened with interest, and felt than an expedition to Africa might just be a viable idea, considering that the *n'yamala* sounded much like the *mokele-mbembe* of the Likouala swamps in the Republic of the Congo to the northwest. However, both men decided that more research was needed before any final decision on an expedition was made, and Powell returned to Gabon in January 1979 to gather more information on the mysterious *n'yamala*. Establishing a base at the Albert Schweitzer Hospital in Lamberene, the intrepid explorer recorded the following information:

> "Albert Schweitzer believed in the sacredness of life, not just human life. Therefore, when the staff of the Albert Schweitzer Hospital at Lamberene (Gabon) learned of the conservation nature of my work, they allowed me to use the hospital as a base. While there, I became friends with a Swiss dentist, who had married a Fang girl from a village a short distance up the Ogowe from the hospital. Accompanied by him, I visited his wife's village, and there became acquainted with the village witch doctor. My dentist friend told me this witch doctor—a man of around seventy—was extremely intelligent, and so I found him to be. Using the dentist as an interpreter, I questioned him. First, I showed him pictures of African animals found in the Gabonese jungles—leopard, gorilla, elephant, hippo, crocodile, etc.—and asked him to identify each one, which he did unerringly. I then showed him a picture of a bear, which does not occur in Gabon. This he could not identify. 'This animal not live around here,'

he said. I then showed him a picture of *Diplodocus* (a brontosaurus-like dinosaur) from a children's book on dinosaurs and asked him if he could recognize it. 'N'yamala,' he answered quite matter-of-factly."

Obiang later told Powell that he had actually observed a *n'yamala* when he was twenty six years of age, which would have been in 1946. Obiang gave his age as 59 in 1979—not 70, as previously reported. At that time, Obiang had camped at a small jungle lake near the Ikoy River where it branched off the N'Gounie. Around 5:00 a.m., Obiang saw a *n'yamala* emerge from the jungle pool to feed on the vegetation. He further stated that the animals would emerge from the rivers and swamps to eat the vegetation between midnight and 5:00 a.m. Powell asked if could visit the location where Obiang had seen the animal, and a river trip was quickly arranged with Powell being accompanied by Obiang, Daniel (the canoeist), and Alec, a carpenter from the Albert Schweitzer Hospital. The group soon arrived at the location where the witch doctor saw the *n'yamala*, and Powell was not disappointed.

The location was exactly as Obiang had described, with a dense rainforest and a small pool about 100 feet across just off the N'Gounie. Unfortunately, the area was literally moving with flies and black ants, making any prospect of a lengthy stay a most unpleasant prospect. With the use of a beer bottle and some line, Powell established the depth of the pool as 18 feet. His African companions, however, refused to spend any time on the water with Powell, lest his depth sounding activities rouse a *n'yamala* that may have been in the pool. Powell later commented that the considerable fear exhibited by Obiang convinced him that the old witchdoctor had indeed seen one of the animals, which he further described as about 33 feet long, with a bulky body as big as an elephant, and a long neck thin neck, a small head, and a powerful, flexible tail.

Although Powell did not observe a *n'yamala*, he continued about 20 miles north on the N'Gounie River and made further enquiries among the Akele people. They were familiar with the name *n'yamala*, but none of the tribe claimed to have ever seen one, and the picture of the *Diplodocus* earlier identified by Michael Obiang did not solicit any recognition—or at least they were unwilling to part with any information to a white outsider. Powell had better luck when he decided to make enquires among the Fang people who lived in a village located midway between the N'Gounie and the O'gowie. Here, three people independently identified the *Diplodocus* as the *n'yamala*—an animal they had each observed in the area. James Powell was on the right track at last.

After carefully considering Powell's findings, Roy Mackal decided that an expedition to equatorial Africa to inquire further about the mystery water beast would be in order. Most of the reports of the *n'yamala* corresponded closely to the *amali*, reported in Cameroon in the 19th century by Trader Horn. This in turn compared favorably with the *jago-nini* and the *m'koo-m'bemboo* of Cameroon of the 1930s. However, many of the areas where the animals were previously observed were now more heavily populated with a corresponding increase in river traffic. As a result, Mackal speculated that the range of *mokele-mbembes* had been greatly reduced in the past 200 years, with the main population now possibly concentrated in the Likouala Region, a 55,000 square mile area of largely unexplored, seasonally inundated swamps located in the north of the People's Republic of the Congo.

In February 1980, Mackal and Powell arrived in Brazzaville, the capital of the People's Republic of the Congo, and then flew north to the town of Impfondo where they met Eugene P. Thomas, who had now been living in the area for 25 years. After settling in at the mission station, Gene Thomas sent out messengers to bring in anyone who had observed a *mokele-mbembe* firsthand. One of the first eyewitnesses to be interviewed was Marien Ikole, a pygmy who lived near the mission station. Originally from the village of Minganga, located in the north near the Ikelemba River, Marien told Mackal and Powell that *mokele-mbembes* were frequently observed in the Ikelemba and Tibeke Rivers and came out to feed mainly at night. Marien also picked out a picture of the *Diplodocus* from a book on dinosaurs as being a good likeness to *mokele-mbembe*.

The second eyewitness was 59-year-old Firman Mosomele. In 1935, when he was 14 years old, Firman saw a *mokele-mbembe* in the Likouala-aux-Herbes River as he was paddling round a bend just south of Epena. He got a good look at a small snake-like head, 6-8 feet of neck, and about six feet of broad back rise above the surface briefly before he made good his escape. The entire animal, according to Firman, was reddish brown in color and was definitely not a snake, a hippo, or any other kind of animal with which he was familiar. Without any prompting from his interviewers, Firman picked up the picture book on dinosaurs and unhesitatingly picked out a picture of *Diplodocus*. "Mokele-mbembe," he declared.

According to some of the informants interviewed by Mackal and Powell, the epicenter of *mokele-mbembe* sightings was to the west of Epena. Villagers from north and south of Impfondo seemed unfamiliar with the animal, or at least no one had seen them for a very long time. The only alternative for Gene Thomas, Mackal, and Powell was to head west to Epena through the *Vassiere*, or "Great Swamp," on foot, as the local flight from Impfondo to Epena was not available. Progress was slow and frequent stops through the mud, sometimes waist deep, were an absolute

Courtesy Dr. Roy P. Mackal / *A Living Dinosaur?*

Dr. Mackal's (right) First Expedition,
with James Powell (left) and Boubaker Ehoula (middle).

necessity for the explorers to replenish their energy in the heat and humidity, particularly as the temperature soared to an almost unbearable 90 degrees Fahrenheit. The pygmies, small in stature but immensely strong and hardy, were easily able to maintain a brisk pace while carrying most of the food, supplies, and heavy backpacks.

After eventually reaching the town of Epena, the team was invited to meet with President Kolonga, the government head of the Epena District. Gene Thomas spoke for the team, explaining the nature of the visit to President Kolonga, and the interest in learning more about *mokele-mbembe*. Whether in jest or deliberate deception, the president casually told the group that *mokele-mbembe* meant "rainbow." Of course, Roy Mackal became a little miffed at this comment and proceeded to inform the president through Gene all that he knew about the suspected dinosaur, including its description, dangerous behavior, herbivorous diet, and places in the river where they have been observed. Astonished at Mackal's knowledge, the president later decided "upon reflection" to send out an appeal for local eyewitnesses to come forward and provide as much information about *mokele-mbembe* as possible.

Within 24 hours, a sizeable room in President Kolonga's comfortable home was full of local hunters and fishermen, waiting patiently to meet with the American who had come from a land so far away from them to learn about an animal they didn't even care about. Nevertheless, they were at least willing to share their knowledge out of courtesy to the president of the region, and to keep within his good graces. In the Congo, tribal rank and social structure were strictly observed and kept. If the government head of the Region de la Epena asked for something—he got it. Mackal and Powell were ready for a long day of interviews and questioning. What would they discover?

At 9:00 a.m. the following morning, President Kolonga welcomed Mackal, Powell, and Thomas back to his home and introduced them to the village elders of Epena region, who were held in very high esteem. As the interviews began, Gene Thomas translated everything from Lingala, the *lingua franca* of the people, into English as Mackal and Powell took notes. The first interviewee was Pascal Moteka, who was formerly from a village located near Lake Tele. He was familiar with the story of at least one *mokele-mbembe* being killed at the lake, and that the stake barrier erected by the pygmies in one of the *molibos* (channels of water connecting the lake to the surrounding swamps) was still there. Pascal further revealed that he still fished in the lake often and still saw *mokele-mbembes* there from time to time, raising their long head-necks out of the water to feed on the vegetation. The necks of the animals he observed were at least two meters in length, and the animals would often wave their heads from side to side and arch their backs out of the water like a buoy. Pascal always observed the animals from a distance from them for fear of being attacked. The second eyewitness was David Manbamlo, a teacher from Impfondo. Two or three years before, he was upstream from Epena in a canoe on the Likouala-aux-Herbes River when a strange creature with a 2-meter long neck and small snake-like head emerged from the water. Although David was badly frightened, he remained still and watched the animal as it raised the upper and breast area of its body out of the water. At that time, the plants and fruits upon which *mokele-mbembes* were reputed to thrive grew in abundance in that location, but have been cut down since. After providing his testimony, David picked out a picture of the *Brontosaurus* (now called *Apatosaurus*) from Mackal's picture book on dinosaurs as being the closest thing to *mokele-mbembe*.

The third witness was Nicolas Mondongo from the village of Bandeko, located near the confluence of the Moudagouma and Likouala-aux-Herbes Rivers. Nicolas stated that his father had seen a *mokele-mbembe* come out of the river and onto a sand bar, where it left dinner plate-sized footprints in the sand, and a trench where the tail dragged along. Years later, when Nicolas was 17 years old, he observed a *mokele-mbembe* for himself in the same part of the river early one morning, and wondered if it was the same animal his father had observed. He had just beached his canoe to hunt some monkeys in the treetops nearby when a violent displacement of water caught his attention. Rising out of the water, Nicolas gaped as he observed a strange animal with a 2-meter neck, bulky hippo-sized body, four sturdy legs and a long flexible tail emerge from the river and remain exposed for a good three minutes. The animal was a good 10 meters (33 feet) in length, gray all over and sported a crest or rooster-like frill at the top of its head. Other reports were vague second or third-hand accounts, so eventually the team headed upriver with David Mambamlo, who showed them the exact spot where he saw his *mokele-mbembe*. No strange or monstrous animals were observed, but later Mackal and Powell had the good fortune to met Colonel Pascal Mouassipposso, the head of military security for the capital city of Brazzaville.

The Colonel originally hailed from the village of Tanga and had observed *mokele-mbembes* in the river on two occasions. A huge, physically imposing man, Colonel Mouassipposso was very matter-of-fact when discussing his own encounter with the much-feared water monster. The first occasion was in 1948, when he and his mother were paddling upstream from Epena—one of the animals crossed the river a mere 30 feet ahead of them. His second encounter occurred that same year, this time downstream from Epena on the Likouala-aux-Herbes River. As he was paddling his canoe in the middle of the river, his vessel suddenly ran aground. Thinking that he had hit a submerged log, the young Mouassipposso was astonished to see the "obstruction" was a *mokele-mbembe* just under the surface of the river. The animal was not interested in a confrontation and moved away from the canoe into deeper water. The colonel also stated that the animal digs caves or lairs in the riverbank, and these sometimes have openings inland.

Finally, an eyewitness testimonial was recorded with Daniel Omoe, an employee of the Ministry of Agriculture. Only six months before Mackal's arrival in the Congo, the inhabitants in and around the Village of Dzeke, some 50 miles downstream from the Epena, regularly observed a *mokele-mbembe* that inhabited a swamp pool just off the main river. As the dry season progressed, a sandbank appeared in the middle of the river, where the people would see the animal enter the river from the jungle almost on a daily basis. Those brave enough to approach the sandbank found fresh, elephant track-sized prints with claw marks, and a 2-meter wide path leading from the water's edge into the swamp.

With time and money running out, Mackal and Powell returned to the United States, tantalized by the amazing reports they had collected. Although Roy Mackal was keen to organize a bigger, more elaborate expedition, James Powell felt that he did not want to return to the Congo, and went back to Texas.

1981

During the planning of his second expedition, Mackal received hundreds of letters worldwide from people asking how they might join a "dinosaur hunt" in Africa. Among the correspondence was a letter from Herman Regusters, an electronics engineer and consultant to the Jet Propulsion Laboratory in California. Regusters offered some impressive technical assistance, such as the use of a portable GPS/Manpac receiver, which could determine the position of the expedition in the Likouala Swamps at any given time via the NAVSTAR and LANDSAT satellites. As the expedition planning progressed, Regusters, it seemed, allowed himself to get carried away with the idea of hunting down possible living dinosaurs in the heart of Africa, and held a press conference at the Los Angeles Museum of Natural History on June 10, 1981. At the gathering, Regusters gave out a press release, which talked about African "alligators," pygmies with blow darts, cannibals, headhunters and the like, all highly inaccurate and grossly misleading.

Eventually, Roy Mackal and Herman Regusters parted company. Roy's disciplined, scientific approach to the expedition and its objectives clashed with Herman's more sensational, media-hungry style to hunting dinosaurs in the heart of Africa. However, Regusters led his own expedition to the Congo and even managed to reach Lake Tele, the source of many *mokele-mbembe* reports.

On the morning of October 27, 1981, Roy Mackal and his team landed at Brazzaville's Maya Maya Airport and were met by Gene Thomas. Mackal's US team included Marie Womack (a photographer from Santa Monica, California), Richard Greenwell (from the University of Arizona and secretary of the International Society of Cryptozoology), and Justin Wilkinson (a geomorphologist and graduate from the University of Chicago). As before, Gene Thomas would join the expedition as the official translator.

Eventually, after dealing with lost luggage, long delays in acquiring government documentation, and mounting hotel bills, the team flew north to Impfondo, where a giant wooden dugout canoe or *pirogue* awaited them. On this second expedition, Mackal's team penetrated the narrow and hazardous Djemba canal into the Tanga River, then into the Likouala-aux-Herbes River and on to Epena—*mokele-mbembe* territory.

From Epena the expedition headed directly for Dzeke, where it was rumored that a *mokele-mbembe* had taken refuge in a place called the Mboueka Pool just off the Tanga River. During the journey to Dzeke, the two expedition dugouts rounded a bend in the river to be met with a large splash, followed by a 10-inch high wave that emanated from a shaded area of riverbank and buffeted the expedition *pirogues*. The pygmies screamed "*Mokele-mbembe, Mokele-mbembe!*" Although the Mackal and his colleagues did not see the animal that had hastily submerged, the pygmies were adamant that they saw the back on the animal as it slipped underwater. What could it have been? The forest elephant that inhabits the area cannot submerge completely. Crocodiles leave a different wake in the water, and hippos are not present in the area as *mokele-mbembes* apparently chase them away or kill them outright. After much poking and prodding with 3 meter-long wooden paddles, the team found that an underwater shelf had allowed the mystery animal some leverage in order to browse on the molombo fruits which grew in the immediate area in abundance. After being startled by the sudden appearance of the expedition canoes, the animal simply stepped off the shelf and sank into deeper water.

Reluctantly, the expedition left the area and set up base camp, established at the village of Kinami. From the base camp, the expedition visited other villages along the Likouala-ax-Herbes River. Although a few eyewitness accounts were provided, a young boy confided to Mackal that he knew of a location on the river where the team would be able to observe a *mokele-mbembe* at a specific time during the day, but he was afraid of being attacked by the village elders. Fear of reprisals from the elders, coupled with mystical and supernatural beliefs surrounding *mokele-mbembe*, made information-gathering difficult, but not all the inhabitants of the Likouala were

COURTESY DR. ROY P. MACKAL / *A Living Dinosaur?*

Dr. Mackal's Second Expedition.

Left to right: Richard Greenwell, Justin Wilkinson, Marien Ikole, Marie T. Womack, Roy P. Mackal, Eugene Thomas, Pascal M'Beke, Marcellin Agnagna, Edouard Dewi, Pascal Botenge. Kneeling: Georges N'dossa, Celestin Kombe.

so intimidated and a few fishermen and hunters spoke openly about the animal, even if some of them had never actually seen one.

Farther downstream at the village of Dzeke, Emmanuel Moungoumela, a great elephant hunter and very well respected in the area, took the explorers to the Mbouekou pool, located in a small area of swamp just off the river where he believed that a *mokele-mbembe* had spent some time, probably browsing on the vegetation there. Intrigued by the reports, Moungoumela discovered a spot he believed the animal had vacated only minutes earlier, as evidenced by droplets of water of the foliage, a great breech made by the animal as it forced its way through the bush, and river grass broken to a height of 1.5 meters as though a heavy tail were swinging from side to side. The hunter followed the trail to the water but did not find any corresponding egress on the other side. Hippos do not live in that area, and the elephants that do will not remain in the river, but exit the water as soon as they have crossed.

During Mackal's research on in the Likouala Region, the Herman Regusters Expedition had made it to Lake Tele where he claimed that he and other members of his expedition made multiple sightings and an alleged encounter. But, four years later during my own expedition to Lake Tele in 1986, my colleagues and I were unable to find a single eyewitness alleged to observe the "long necked member" at the lake. In spite of the many cameras that the Regusters expedition had with them, not one single photograph or piece of film was produced to corroborate the multiple sightings that allegedly occurred during their 32 days at Lake Tele. To be fair, the Regusters expedition was in the right place at the right time to have observed a *mokele-mbembe*, so Herman and Kia Regusters may have had a close encounter with one of the animals.

MARCELLIN AGNAGNA

One of the Congolese members of the Roy Mackal expedition in 1981 was Marcellin Agnagna, a zoologist based at the Brazzaville Zoological Park. Encouraged by the reports of the eyewitnesses who claimed to have observed *mokele-mbembe*, Agnagna sought the permission of the Brazzaville government to conduct his own expedition to Lake Tele. On May 1, 1983, Agnagna was filming a troupe of monkeys in the trees surrounding the lake when he was alerted by the shouts of one of his companions. Standing by the edge of the lake, Jean Charles Dinkoumbu from the village of Boha was pointing to something out in the water. At first Agnagna's view was obscured by overhanging foliage, but seconds later, a strange animal came into view with a long neck, small head, and broad back. The zoologist's attempt to film the creature with a Minolta XL-42 movie camera was apparently disrupted by the shock and fear exhibited by Dinkoumbu and other members of the expedition at the sight of the animal. The film in the movie camera had almost all been exposed by this time, and the lens was set on the macro position, which was useless for long-range filming.

Agnagna later reported that the frontal part of animal's face and head was brown, while the neck and back was black and shone in the sunlight. The animal moved around as if to determine the source of the noise as Dinkoumbu continued to shout with fear, then gradually submerged with just the long head-neck above the surface before eventually slipping underwater. Agnagna's film was later developed in a French laboratory, but revealed nothing of interest in connection with the alleged encounter with a *mokele-mbembe*.

At the time of writing this book, I discussed Marcellin Agnagna's sighting with Roy Mackal. Not only was the sighting too late in the year for a *mokele-mbembe* due to their reproductive and hibernation cycle (more on this in Chapter 8), but Roy and I felt that Marcellin had actually observed a very large specimen of fresh water turtle, *Trionyx triunguis*, which can attain a shell width of two meters, with even larger specimens being reported in the lakes and rivers from time to time. Again, as with the Regusters expedition, I was unable to locate any of the alleged witnesses that could corroborate Agnagna's observations, during my own visit to the village of Boha in 1986.

Operation Congo

In May 1985, Brighton University was to host the annual conference of the International Society of Cryptozoology. Attending would be Richard Greenwell, Roy Mackal, and the great Bernard Heuvelmans. Excited by the possibility of meeting these unique men, Mark Rothermel, Cheryl Pirelli (Mark's knockout girlfriend), and I drove south to the scenic west coast and attended every lecture that day, including one on the existence of a hitherto unknown species of giant octopus, by Roy Mackal. Also in attendance was Tim Dinsdale, an aeronautical engineer from Hertfordshire who had devoted many years of research at Loch Ness. After the conference had ended, Mark, Cheryl, and I made our way to the picturesque seafront hotel in Brighton to pick Mackal's brains on his research in Africa. At that time, Roy was working on a book on *mokele-mbembe* and was reticent to part with any unpublished information. Greenwell was pleasant enough and we found Bernard Heuvelmans to be a patient, kindly old man who answered our questions courteously. Unfortunately we did not meet Tim Dinsdale, but I did have several telephone conversations with him prior to our departure for Africa. Tim sent me copies of his books, *Loch Ness Monster*, and *The Leviathans*. Just a week before leaving Britain, Tim also sent me a shiny cap badge of the Royal Ghurka Regiment. He had carried it throughout his many years at Loch Ness as a good luck charm, and now he was passing it onto me. I was speechless with gratitude. What could I say?

It was time to acquire our visas from the Congolese Consulate in Paris. Just one week before I was due to sail for France from the white cliffs of Dover, I carelessly fractured my left ankle when sprinting across a London street to beat the oncoming traffic. My hopes of going to Africa were suddenly dashed as I headed off to the hospital in a taxi. Two hours later I hobbled out of Emergency on crutches with my left ankle encased in a plaster cast that reached up to my knee. Forever the optimist, Mark didn't feel at all put off by my sudden misfortune and thought that as we still had a month to go before leaving the country, perhaps my fractured ankle would be healed before then. After he drove me down to Dover, I limped onto the ferry and found a seat before heading off into the choppy waters of the English Channel, bound for Calais. As the ship progressed out into the channel, the sea became increasingly turbulent. Yet in spite of the roller coaster journey, I felt quite hungry and made my way to the restaurant and ordered a plate of fish and chips with a piping hot mug of tea to swill it all down. As I tucked into my dinner, numerous green-faced passengers battling sea sickness gave me odd glances, wondering what my secret was, being able to eat while they could barely think of food without being violently sick.

On the journey I found myself in the company on Nigel Harding and Edna Harris. Nigel was an author who wrote about travel and culture. He was on his was to France to pick up some books he couldn't obtain in England. Edna was a sprightly 78-year old Englishwoman who had married a Frenchman and lived in Florence. During World War II, Edna was recruited by British Intelligence and parachuted behind enemy lines to work with the French Resistance. The Gestapo knew she was in the country, but could never find her. After the war, she married a Frenchman and started her own business. As we talked about our various reasons for visiting France, Nigel took keen interest in my forthcoming expedition and even offered to accompany me to Paris to assist with securing my visa, which I gratefully accepted. After arriving in Calais, we caught the train to Paris and arrived three hours later. At the station we hailed a taxi cab for the journey to the Congolese Consulate. Our driver was a diminutive, elderly French woman with an equally tiny French poodle that sat in the front passenger seat on a plump embroidered cushion. Before we could settle into the back seat, Granny took off like a rocket, weaving in and out of the traffic with the skill of a Le Mans racing driver. Nigel and I gaped wide-eyed at the traffic on both sides of the road as we raced through the narrow streets, wondering if we would ever get to the consulate alive. Thankfully we arrived at our destination in one piece and paid Granny before wobbling our way into the consulate on our jellied legs. The building was almost bare of any furniture. The carpets were threadbare, the light bulbs were bare and unshaded, and the toilets didn't have any tissue paper or soap. The consulate staff was rude and surly, and not in the least interested in the letter I showed them from the Minister of Water and Forests, inviting us to the Congo for our expedition. One consulate employee, a charming French woman, took pity on my dilemma and advised me to return to her office after lunch, when she would have my visa ready. Nigel and I took her advice and went to lunch where he picked my brains on the expedition in general and *mokele-mbembe* in particular. An hour later we returned to the consulate where my passport was waiting with a three-month visa within for the People's Republic of the Congo. Thankful for the effort that the French lady had put in on my behalf, Nigel helped me to select a bouquet of flowers, which we presented to her in gratitude for her kind assistance. The train ride back to Calais and a simple dinner completed my adventure in France. Nigel and I shook hands, and parted company. He went on his way to hunt down some books and I returned to England with my newly acquired visa and three visa application forms for Mark, Jonathan, and Joe. I was grateful to have met Nigel, whose fluency in the French language went a long way to ensuring my success in Paris. It would have taken far longer for me to have achieved the same simple task with my own schoolboy French.

On November 17, 1985, I departed from Great Britain, bound for the Congo. Our team of four explorers included Mark Rothermel from the county of Essex, who had previously explored some of the tributaries of the mighty Amazon three years before; Jonathan Walls, a childhood friend of Mark's and a history teacher, now our official French translator for the expedition; Joe Della-Porta, an army sergeant and expedition medic; and myself, team leader and cryptozoologist. Mark, Jonathan, and Joe flew from London, England, to Brazzaville via Aeroflot, the state airline of the Soviet Union. I had opted for a two-week journey by ship to Pointe Noire on the Congo's Atlantic coast with all our expedition equipment onboard the *Sokoto*, a 5,000-ton cargo vessel, courtesy of Elder Dempster Shipping Company. The journey helped me to gradually acclimatize to the African heat as we stopped in Gambia and Nigeria prior to arriving at my destination, the deep-water port of Pointe Noire in the Congo.

The expedition had taken 18 months to plan, which included months of correspondence (all in French) with the Congolese government in Brazzaville, including Marcellin Agnagna, who had agreed to participate in a joint British-Congolese venture. We had gleaned almost all our knowledge about *mokele-mbembe* and its habitat—the

Likouala Swamps—from the Mackal, Regusters, and Agnagna expedition reports, which were published in the journal of the International Society of Cryptozoology.

After arriving at Pointe Noire, our expedition equipment, safely packed in four wooden crates, was whisked away into custom bond before I could retrieve them. Two days later I joined the rest of the team in Brazzaville and met with Marcellin Agnagna. To our utter disappointment, no official documents had been prepared prior to our arrival, and we had to attend meetings with the Secretary General Francois Ntsiba and Henri Djombo, the Minister for Water and Forests to establish our expedition objectives and area of interest. Had our plans not been clearly laid out during the months of correspondence with the same ministers? Why did we have to go through every little detail again? Our stay in Brazzaville lasted for eight frustrating weeks, during which time we fought through mountains of red tape, and struggled against an incredibly corrupt and completely incompetent bureaucratic system. The Congolese, it seemed, were prepared to do absolutely nothing unless they saw the whites of your eyes.

Like every other visitor to the Congo, our passports had been held by the Ministry for Internal Security, which issued us with a single white piece of paper with our details printed on it and a passport photograph of the bearer attached. The clerk in charge of all foreign passports had "lost" our documents, but was willing to "find" them again if we paid his café bill. Needless to say, we refused. Marcellin Agnagna did his best to help with the seemingly insurmountable corruption and unbelievable government incompetence, and slowly we began to make progress. We eventually met with Pastor Eugene Thomas, who was shocked at the length of time everything was taking, and did his very best along with Marcellin Agnagna to move things along. Our equipment was still being held in Pointe Noire, and the only way to retrieve everything was to return to the coast and meet with the chief customs official.

Our money was beginning to run low, which necessitated Mark sending a request for more funds from his father, Rolf, a prominent London architect, and our main expedition sponsor. After the money arrived, I set off for Pointe Noire by train with Djoni Jose Bourges, the Chief Game Control Officer for the Congo, who would join us on our expedition with Marcellin and constantly address me as "Mr. Bill," which sounded like "Meester Beel" in his French-African accent.

As we sat in the First Class compartment of the fast moving diesel train, we discussed the expedition itinerary, including, of course, *mokele-mbembe*. Jose was convinced the animal existed and felt that there was certainly enough evidence to justify more funding for research by his own government. Looking out of our compartment window as we sped south to Pointe Noire, I thought about the 332 miles of railway line that linked Brazzaville to the picturesque Atlantic coastline, known as the Congo-Ocean Railway. Built by the French between 1924 to 1934 with forced labor recruited mainly from Chad and the Central African Republic, it is estimated that close to 17,000 Africans died during its construction. The railway, which the Congolese are proud of in spite of the human cost to build, pushes through the dense Mayombe rainforest where elephants are still present today. Each time we stopped at some small town or village, women and children would crowd at the windows selling fruit, cooked meat, fish, and fresh sandwiches.

The situation in Pointe Noire was little better than in Brazzaville. It took days to see the *Chef de Customs*, who took little interest in our dilemma, in spite of the fact that I had a letter from the Minister of Finance exempting us from paying any fees for the weeks that our equipment and supplies were held in bond. Eventually, with much haggling from Jose, our equipment was released. On going through our crates with yet another petty bureaucrat present, I discovered that one of our cartons had been broken into and a large amount of clothing and personal kit, including several expensive hunting knives, had been stolen. I demanded to know just how this could happen when our crates were supposed to be in "protective" custody. The bureaucrat shrugged and the customs officer present just didn't care and felt he did not need to explain himself. In the Congo it seemed that if the white man didn't like it, the white man could leave. "Well," I thought, "welcome to Africa!"

After returning to Brazzaville with our equipment and supplies, Jonathan typed up a document outlining our expedition objectives (again), including brief biographies on all four Britons. This done, we set out to purchase clothing and supplies for our two Congolese colleagues, who we also had to pay for the duration of the expedition. In all, at least $5,000 (US) went to providing food, clothing, backpacks, tents, sleeping bags, cameras, film, and other items, including all travel expenses for Marcellin and Jose. Another fee went to the two ministries that provided our paperwork, allowing us to travel outside Brazzaville without being arrested. At least $7,000 went toward these additional expenses and we hadn't even left Brazzaville.

With all the delays and frustrations, we had celebrated Christmas, the new year of 1986, and my 28th birthday on January 7. The extended stay in Brazzaville did at least allow us to grow more accustomed to the country, the culture, and make some good friends with the locals. We also took the opportunity to pump information on *mokele-mbembe* from Marcellin, Jose, Pastor Thomas, and even Colonel Pascal Mouassipposso, who invited us to his office to discuss our expedition. A huge, imposing man in his 40s, the colonel outlined his own plans for a full military expedition to Lake Tele. There, he would post soldiers all around the lake and wait for a *mokele-mbembe* to appear. Whether his plan was to kill or merely film the animal was never made clear, but his own two encounters with *mokele-mbembes* in the river near his home village of Itanga were proof enough that finding a specimen and presenting it to the world was his ultimate goal. In February 1986, we bade farewell to Brazzaville, boarded the small twin-engine Fokker 128 passenger jet and flew to Impfondo, the launching ground for our expedition. Marcellin and Jose would follow.

Having arrived in Impfondo, a government truck delivered us to the mission station run by Gene and Sandy Thomas, who would be arriving in another week with a mission team around the same time as Marcellin and Jose (who were delayed due to yet another bureaucratic problem in Brazzaville). Once our tents had been raised, we settled down for the night after a supper of fried eggs and luncheon meat, washed down with hot coffee. For hours, pygmy adults and children stood in a large group, watching us intently as we established our camp. It seemed they were just as curious about us as we were of them. After yet another week of kicking our heels, another meeting followed with the *Commissar Politique*, the political head of the Impfondo Region. After examining our papers, he decided to let us proceed with our expedition. Marcellin and Jose arrived without a small 2 HP outboard that we needed for our inflatable boat. After another frustrating wait while Marcellin tried to find someone—anyone—in Brazzaville who could be bothered to put the engine on the plane and send it to Brazzaville, we gave up any hope of seeing it and decided to proceed without it. After two weeks in Impfondo, we were set to catch the small twin-engined DeHavilland Otter and fly over the *Vassierre*, or great swamp, that Roy Mackal and his team were forced to trek through four years before. The flight took only 20 minutes, which was preferable to slogging through the mud for two days.

It was during my time at the mission station that I began to question why I even bothered to come to Africa. In London, I had a few friends, a steady job, and a place to live. But my peace of mind and direction in life had been confused and distorted with my careless dabbling in the occult. I had developed an intense fear of the dark and horrific nightmares plagued my sleep. In short, I felt haunted and confused about where my life was heading. All the problems that my colleagues and I were facing in Africa only compounded my own sense of misery and hopelessness. I wanted answers and direction, but where to find them? Spiritualism, tarot cards, and associating with those involved in satanic arts only made matters worse. As the sun dipped down behind the dense forest behind the mission station, I found myself opening up to Gene Thomas, a powerful preacher and loyal soldier for Christ. He listened patiently and then recounted his own story, about how as a young man he was a thief in a gang and gave his own Christian parents nothing but grief. One night just to keep the peace, he accompanied them to church and found himself so profoundly convicted of his sinful life that he literally ran down the isle to give his life to Christ. His joyful parents joined him at the altar, giving thanks to the Creator of the universe for bringing their son to the Cross. There was much in his testimony that I could relate to. But my own problems ran deep and dark. The occult was a dreadful thing to get involved with, and its black tentacles had wrapped themselves so strongly around my soul that no escape seemed possible. Night after night I questioned Gene and poured out my heart, relating my problems and fears and my confusion to him. Patiently and with deep concern for my situation, the missionary from Ohio showed me answers to my problems from the pages of the Holy Bible. I was shocked to see answers—real answers to my problems—leap out of the pages at me. Gene fixed his deep brown eyes on me and spoke with genuine conviction and concern. "Bill, you have opened up your life to Satan and his lies. He won't let you go without a fight. The only way out is for you to give your life to Christ." Deep down in my heart I knew he was right. Although his days were filled with mission business, this extraordinary man of God took time to spend with me talking into the wee small hours, going over my problems, showing me answers in the Bible and ending each meeting with a prayer. Every meeting opened my eyes to the evils that I had exposed myself to, but the black tentacles of the occult seemed to tighten their grip on my soul.

One night as I lay in my bed unable to sleep, the night seemed eerily silent. Why were there no barking dogs, no shrill cry of the leopard in the forest, and no chattering bats in the trees? Even the insects were silent. While I was slowly drifting towards sleep, something caused me to suddenly snap awake. There was something else in the room with me. At first I though that Mark or Jonathan had slipped into the room to play a practical joke on me, which they often did. But this was different. Although I couldn't see anything in the darkened concrete room, a presence, real but unseen began to manifest. Within a minute, perhaps less, a dark and thoroughly evil entity began to fill the room. I tried to call out to Mark, Jonathan, and Joe, but my voice died in my throat. I tried to sit up and grab a box of matches by my bed to light the little kerosene lantern, but I was completely paralyzed. Steadily the evil presence approached my bed. I was fully awake, yet completely immobile, soaked with sweat and beginning to go out of my mind with terror. "What are you?" I cried out in my mind. "What do you want with me?"

As if to answer me, the most unbearable pressure suddenly began to press down on my chest, shoulders and arms. The crushing pressure squeezed the air out of my lungs as I began to lapse into unconsciousness. I fought to stay conscious but a dark abyss seemed to open before me as I began to black out. Then the words of Gene Thomas sprang into my mind. "Only Christ can set you free."

With every last vestige of conscious will, I cried out in my mind to the savior that the Christians worshipped. "Jesus, help me!" Again the pressure continued. "Jesus, help me!" The black, unfathomable darkness began to slip away. "Jesus, help me!" The pressure began to subside. "Jesus, help me!" The suffocating, evil presence in the room recoiled, and then vanished. The night sounds of the jungle started up again as I lay in my bed, terrified, waiting for the first light of dawn to break the black night of the Congo. The following morning I made my way to the mission station and told Gene what had happened. The grave look of concern on his face spoke volumes. Now was the time to decide if I wanted to break away from occultism or remain living in my current spiritual and mental condition. Still I hesitated, feeling more uncertain than ever. I did not share my encounter with evil with my fellow explorers, lest they conclude that I was either losing my marbles or smoking some illegal substance. Later we decided to eat dinner at a local bar where the music was loud, the beer was awful, and the locals expected us to buy them drinks all night. Then it happened.

Suddenly I found the whole situation unbearable. Desperate to get away from my friends, the bar, the locals and the blaring music, I made the excuse of feeling tired and slowly walked into the darkness towards the non-descript concrete building where evil was waiting for me. Each step I took felt incredibly heavy. What was wrong with me? Every bad thing I ever did, every negative word I ever spoke, every rash decision and dishonest act I had committed in my 28 years just seemed to weigh me down. I felt like Atlas supporting the world on his shoulders. But I was no Greek god. I was a mere mortal, lost, confused, and tired of the life that was filled with darkness, fear, confusion, and with no moral guidance. I knew I had reached a crossroads in life. The question was which road should I take?

I pushed open the front door and stood there, listening for anything that might be lurking in the shadows. All I could here was my own breathing and the familiar noises of the night outside. I entered my room, lit the kerosene lantern and sat on the floor by my bed feeling completely dejected. I pondered on the Thomas family and the other missionaries I had encountered in Africa. Why were they so happy? Why did they seem at peace? What spirit burned brightly within them that motivated them so much to proclaim the "Good News" of Jesus Christ in the face of the many dangers they faced in the heart of Africa? I decided to give their God just one chance to prove that He was real.

Kneeling by my bed, I called out to Jesus and asked Him in plain language, to show me that he was for real, to come into my life and wash away the confusion, the unhappiness, and the dark fears that gripped my soul. There was no burst of heavenly glory and no angelic choir. As I poured out the dark contents my heart to God, a peaceful presence began to fill the room. That night, I climbed into my bed and enjoyed the most restful night's sleep I had known in many years.

The following morning I felt like a vastly different person. My mind, my heart, and even my very soul seemed to be suffused by an incredible peace. There was joy in my heart and a spring in my step. Even the world itself seemed so different, as if I was looking at everything through new eyes with a new understanding of everything around me. This was no mere religious "conversion," no turning over a new leaf by my own strength. The change was so great that I could barely take in it. This was truly a new birth and one that was aptly described in the New Testament:

> "Therefore, if anyone is in Christ he is a new creation; the old has gone, the new has come!" *2 CORINTHIANS 5:17*

Now, by the indwelling presence of the Holy Spirit, my eyes were opened. Now I truly understood. Now I was plugged into the "source" that is the Creator of the universe through his only begotten son, Jesus Christ. This was for real! Now I understood why just after the crucifixion, the frightened group of apostles had been transformed by the risen Christ into fearless ambassadors of Heaven, who preached the Gospel with power and went gladly to cruel deaths. Now I knew why the missionaries went fearlessly into the darkest jungles to take the Gospel to the hidden tribes in the face of indescribable dangers. No matter what happened to them, they would be safe in the arms of their loving savior. Mark, Jonathan, and Joe all noticed the difference. They too knew that something remarkable had happened to me but had no "rational" explanation. From a religious perspective we were an odd group. Mark was a top student in religious education in high school, yet became an atheist. Jonathan became an agnostic after he was kicked out of church by an overzealous, intolerant elder who didn't like tough questions about the Bible. Joe fell away from the Roman Catholic Church because he disagreed with the teachings of some of the priests who tutored him. And there was me, a former occultist and now a born again Christian.

The missionaries were overjoyed with my coming to the Lord. Gene became my spiritual father and Sandy gave me her own NIV Bible, which I still read today. Although Joe recognized that something remarkable had happened to me, Jonathan and Mark remained skeptical and decided one night to really give Gene a grilling on theology. They fired off dozens of tough questions and tried to point out the alleged discrepancies and contradictions in the Bible. Gene answered every question, every point, and every objection head on. With almost unbelievable patience, he opened his Bible and answered every question. If he didn't have an answer to a question, he would be honest and simply say, "I don't know." At the end of the night, Jonathan and Mark left the missionary with a renewed respect for him.

Two days later we were ready to depart for Epena. A small twin engine De Havilland Otter carried us over the *Vasierre* on what was literally a 20-minute hop. We looked down at the vast stretch of dense swamp and forest below us, grateful that we didn't have to slog our way through it like Roy Mackal four years before. We landed on a small bumpy grass strip with no facilities of any kind.

Epena was a small town on the banks of the Likouala-aux-Herbes River. We had at last made it into the target area, almost four months after leaving England. A brick hut was allocated for our "comfort" and we were summoned to see the local police chief who examined our papers, questioned us on the reason for our visit (as if he didn't already know), and how long we would be staying. Two days later, we visited with President Kolonga, the official *Chef de la Region*. Unlike the fanfare of Roy Mackal's visit in 1981 before, the president did not keep us long, did not summon any *mokele-mbembe* eyewitnesses, did not offer us any refreshments, and paid little attention to Marcellin and Jose, whom, we were assured by them, were "big men" in Epena. So much for their alleged reputation. After three days of exploring the area around Epena, we were able to purchase fuel for our river guide who had arrived the day before from the village of Dzeke with a large wooden *pirogue*, powered by a 20hp outboard. At least he was on time. On a beautiful bright sunny morning, we were on our way to Dzeke where the great elephant hunter Emmanuel Moungoumela lived. What would we find?

The journey to Dzeke was not without brief moments of excitement. Twice we hit submerged logjams that almost capsized our *pirogue*. We passed a few *mokele-mbembe* encounter sites, but for some reason Marcellin was reluctant to stop and examine these for signs of activity that would indicate that the animals might still be around. We reached Dzeke by nightfall and a village elder ushered us to an unfurnished brick house that would be our home for the next two weeks. We set up our tents on the dirt floor inside the structure and slept on our thin foam mattresses. The following morning the *Comite du Village* was briefed on our mission by Marcellin and Jose. Not surprisingly, they were reluctant to talk about *mokele-mbembe* and added no one had seen the animal recently. We were free to explore the area under Marcellin's supervision, which we accepted. Later, we had better luck with Emmanuel Moungoumela, but the great hunter had withered to a mere shadow of his former self.

Shortly after Roy Mackal had left the Congo, Moungoumela became ill with malaria, complicated with dysentery, which necessitated a year's stay for treatment in Brazzaville's University Hospital. The pygmies firmly believed that his illness

A Congo Pirogue,
Borrowed from the Catholic Mission
in Epena (Congo-Brazzaville)

was punishment for his willingness to help the Mackal expedition with information on *mokele-mbembe*. As we talked about his findings at the Mbouekou pool in 1981, Moungoumela casually mentioned that he had even observed *mokele-mbembes* on no less than three occasions since Roy Mackal had been there. His first observation had been in the Sangha River, where he was tracking elephants. He explained:

> "The animal kept its body in the water, but its neck—which was very long—stretched up to the molombo fruits which it ate, along with the leaves from the trees there. The head was like a snake. I stayed there for a long time watching the animal until it was dark. I have heard stories about the animal from those who fear it, but I don't think it is any different from the hippo or the elephant. It is just an animal that lives here."

On his second encounter:

> "I was fishing a lot in the Likouala-aux-Herbes River and saw the animal again twice. It did the same thing. It kept its body in the water while it ate the fruits and leaves. I stayed there and watched the animal until it went down in the water. I saw it again in a different place on the river a while later but it might have been a different animal."

Moungoumela was unable to tell the time and as a result did not wear a watch. But he was able to establish the fact that his observations of the animals covered several hours, giving him ample time to study them closely to be absolutely certain that he was not mistaking a known animal such as a hippo, elephant, or even a large snake for something out of the ordinary.

Later, he took us to the small Mbouekou pool which at least one *mokele-mbembe* had been known to inhabit from time to time. Even two years later, the breech made by a large animal that pushed through the bush, including broken branches and the distinct remainders of its footprints in the ground, now fading with age, leading into the swamp and then into the river. It was a perfect hiding spot for a shy, elusive water animal that was close to the river but not far from the village. The area remained undisturbed by the villagers who regarded the pool as a taboo area. Given the fact that both Roy Mackal and Marcellin Agnagna had examined the pool in 1981 and 1983, it is possible that the location was a regular haunt for at least one *mokele-mbembe*, particularly if its food supply was plentiful.

As our expedition progressed, Marcellin's behavior became more bizarre. He became more reluctant to discuss his own *mokele-mbembe* observations, stopped us from seeking out other eyewitnesses, and became more hostile towards us as the expedition progressed, even objecting to our insect and fish collecting activities. News reached us of a small village near a lake, over a day's walk south of Dzeke, where the inhabitants were familiar with *mokele-mbembe*. With the permission of the elders, we prepared to venture forth into deep rainforest with Emmanuel Moungoumela, his son, Emmanuel "the Young," and two other village hunters. Every time we stopped to rest and drink from our canteens we found ourselves under assault from swarms of tiny sweat bees that would attempt to crawl into our ears and noses.

By noon the following day, we reached Lake M'Boukou, which was perhaps a half-mile in diameter, with dense forest on its northern shore. The village itself was a small mixed pygmy and Bantu settlement, presided over by four sisters, all of them witches. Once again, our mission was explained by Marcellin, and permission was granted to explore the area. Emmanuel Moungoumela did not make camp with the rest of us, but built a small lean-to outside the village and remained there during our four-day stay. We later learned that the great hunter did not like pygmies and refused to associate with them. Marcellin went even further by declaring that pygmies were "not human beings." This attitude shocked us, considering how the Congolese often complained about the crimes committed against them by their former colonial masters. Yet the appalling discrimination and mistreatment by the African against the shy and diminutive pygmy peoples was inexcusable. Years later, I watched a nature show on television featuring a British scientist who studied chimpanzees. "Chimps *are* people," he insisted. How ironic that a British scientist insisted primates were human beings, but an African scientist insisted that pygmies were *not* human beings! Charles Darwin has a lot to answer for. On the third day of our stay in the village, we at last gathered our gear, climbed into a large, leaky *pirogue* and headed off across the lake, which was apparently home to a monstrous crocodile. Jose, who could not swim, asked me to rescue him if our overloaded *pirogue* capsized. As crocodile wrestling was not on my résumé, I didn't answer him. After crossing a large expanse of savannah, we were again enveloped in deep forest. After another hour of trekking, we came to a small river where we made camp for the evening. The river, we were told, was sometimes used by *mokele-mbembes* to move from one area to another. The area was certainly tranquil and remote enough to entice our elusive water monster to spend some time to browse without being disturbed.

Young Emmanuel recalled that about a year previously, he was camped at the very spot where we were resting after a long day hunting monkeys. Early in the wee small hours he was awakened by a tremendous splash as some huge animal crashed down a steep bank into the river. He listened intently, waiting for the animal to exit nearby. However, the animal did not climb ashore, but remained in the river. He was sure that the creature of the night was not an elephant or a hippo, as an elephant would have left the river shortly afterwards, and hippos are not found in that location. The following morning, Emmanuel examined the large breech that the animal had made by pushing through the forest to reach the river, but he was unable to find any corresponding egress that would almost certainly have suggested that the animal was a forest elephant. He believed that the water-loving creature of the night was a *mokele-mbembe*. After a full 24-hour exploration of the river and the immediate environment, we started the long trek back to our base camp at the village. Shortly after our arrival one of the villagers casually mentioned that a deep swamp pool located behind the village was home to a strange animal with a long, thin neck. Without wasting any time, we grabbed our cameras and headed for the location. After only five minutes, we found ourselves sinking into knee-high mud. The deep swamp pool lay before us, and Emmanuel Moungoumela was able to hop onto a fallen tree located directly in front of the pool. Without hesitation, he began to thrust his ten-foot-long elephant spear into the pool, an action that encouraged his son and a couple of curious onlookers from the villagers to beat a hasty retreat from the area. If an enraged dinosaur suddenly exploded from the pool to confront its tormentor, we would have had little chance of a quick escape in the knee-high mud. But, once again, there was no large animal, known or unknown, inhabiting the swamp pool. We headed back to the village to clean up, cook dinner, and prepare for the trek back to Dzeke the next morning.

As we continued to amass interesting reports about *mokele-mbembe* and another strange animal (a rhino-like creature that lived in the river and killed elephants), collect insects and fish, and explore the forest and patches of swamp in the immediate area, *mokele-mbembe* remained elusive. We left Dzeke after a two-week stay, during which time we made numerous friends who showed up to bid us a fond farewell. The large motorized *pirogue* that whisked us from Epena to Dzeke was long gone, and we were left to paddle our way to Boha, which was a pleasant 90 minutes north on the Likouala-aux-herbes River. Although the inhabitants of Boha were just as curious about us as the people of Dzeke, they were less friendly and certainly less welcoming.

We were quickly billeted in another brick house complete with tables and chairs, including wood framed beds with sponge mattresses. Stretching out our sleeping

The Operation Congo Team

Josie Bourges Filling the Expedition Canteens (Likouala 1986)

bags on these, snug within the canopy of our mosquito nets, we enjoyed the most comfortable nights sleep in weeks. The following morning we met with the village chief and the *Comite du Village*, who learned of our desire to visit Lake Tele.

The mood was somber, mainly due to the fact that the Regusters expedition of 1981 had treated the chief with disrespect, and promises of money and gifts for the villager's help in reaching Lake Tele were not fulfilled. This only made it harder for us to win the trust of the villagers, which again necessitated another week of sitting around while Marcellin and Jose negotiated our passage to the lake.

Eventually, the chief sent word to us that he had a dream that he was hunting at Lake Tele and had killed an antelope there. This was apparently a good omen and a sign that we could visit Lake Tele as planned. Two days later, drums were beaten outside the chief's hut and a bowl of palm wine was passed around as we each took a sip. With payment of 100,000cfa (roughly $200 USD) made to the elders, we marched into the deep, dark forest with six Boha villagers, including a tall slender giant of a man called Ebitas, and a short muscular soccer player affectionately known as Pelé, named after the great Brazilian maestro of the world's most passionate sport.

After two days of foot slogging through what seemed like an endless sea of green, we eventually reached a clearing in the forest—a resting place just three kilometers from Lake Tele. The weeks of living in the bush, drinking chlorinated swamp water, and eating mainly African food, soon took its toll on our health. Both Jonathan and I came down with bouts of dysentery and intermittent episodes of fever. Skin rashes plagued us and our energy levels were down. Joe and Mark fared better and were chosen to visit three small sacred lakes in the forest where *mokele-mbembes* apparently spent a good deal of time, using the larger Lake Tele to feed and move around. Alone with only two Boha villagers for company, I spent my time catching up with my notes and resting. Ebitas found a wild lemon tree and eagerly collected a bag full of its fruit. With these, we squeezed fresh lemons into boiling water, added a couple of Saccharin pills and enjoyed a hot lemon drink to supplement our dinner of wild antelope and cassava.

A 24-hour vigil at the three small lakes produced negative results, and a day later we had packed up our camp and began the delicate task of negotiating the last three very swampy kilometers to Lake Tele. Jumping from one exposed tree root to another proved to be a difficult task, especially with heavy backpacks. Mark had made some rudimentary wooden staffs and these proved to be invaluable in helping us to maintain our balance while negotiating the swamp. After three back-breaking hours of sweat-drenched effort, we arrived at the lake and looked out on this little inland sea with a feeling of accomplishment. We had made it! Only a few months before, we were all living in foggy, cold England, and now here we were in one of the world's most mysterious places in the very heart of Africa.

Over time, a thick springy surface made up of dead leaves, vegetation and exposed roots had formed around the southern area of the lake, allowing us to pitch our tents on a far less uncomfortable surface than we were used to. Marcellin secured his tent on a wooden platform at the edge of the lake and continued to sulk, becoming more withdrawn from the group. After supper that night, I asked Marcellin again about his *mokele-mbembe* observation at the lake three years before. He suddenly became agitated and accused me of being "ignorant to question him." The following day he showed Mark the spot out in the lake where he had seen the monster. Later we plumbed the depths and found the lake to be up to a mere 3 meters deep in most places—too shallow to allow the alleged five meters of visible, hippo-sized *mokele-mbembe* to submerge in the manner that he had described. During our time at Boha village, we could not find any of the witnesses from 1983 who could confirm Marcellin's story.

Jose too did not seem keen to elaborate on the alleged multiple sightings of a *mokele-mbembe* by various members of the Regusters expedition in 1981. Regusters and his wife, Kia, may well have encountered one of the animals for themselves when exploring the lake in their inflatable raft, but the lack of other eyewitnesses to these reports was unsettling. On our daily excursions around the lake, we saw plenty of birds, monkeys, gorillas, chimpanzees, crocodiles, pythons, and large monitor lizards, but no large monster. Had we invested all our time, energy and money into a wild goose chase? Did *mokele-mbembe* even exist at all?

On the fourth day the team prepared to paddle for three hours to the northern end of the lake where a series of *molibos* (water channels) merged with the swamps. There too was the location where the killing of a *mokele-mbembe* took place some 26 years before. Too sick to participate in this most important part of our expedition, I lay in my tent, too weak and feverish to go with them. What would they discover?

As the sun began to set beneath the jungle wall, the group returned. Did they actually see the legendary water monster—or at least find evidence that it existed at all? Later, in the privacy of our tent, Mark explained to me what they had found during their exploration of the mysterious northern shore.

As the team slowly approached the northern side of the lake, the fear and agitation of the Boha villagers grew by the second. They simply did not want to spent any time at all exploring the *molibos*, but we had paid them to take us there and Marcellin and Jose were representing the Congolese government. The *molibos* proved to be up to five meters deep, and merged well into the swamps, and connecting to the Bai River to the northwest. Ebitas mentioned that the animals would migrate from the river through the channels in the swamps to the lake and back again. There was no evidence of any current pygmy settlement at the lake, and the wooden stakes that had been erected to block access to the lake from the Oume *molibo* had either rotted away or had been removed to allow Ebitas and others to fish in the area. But the poking and prodding by Mark, Joe, and Marcellin in and around this particular *molibo* where a *mokele-mbembe* had been speared to death 26 years before, almost caused our guides to lapse into complete hysteria. Mark and Joe sensed that the situation was becoming potentially dangerous and posed a threat to their personal safety. With great reluctance, the team called off their exploration of the area and began the long paddle back to the camp.

After a dinner of wild forest pig, cassava and our hot lemon drink, Ebitas confided that he regularly fished at the lake, had seen *mokele-mbembes* on numerous occasions and confirmed that at least two of the animals still lived there but did not want his next observation to be his last, hence his nervousness when approaching the northern side of the lake. Marcellin's bizarre behavior worsened as he argued and fretted over mere trivialities. He boasted that he held a brown belt in Judo, and offered to teach any one of us a lesson if we felt up to taking him on. Mark, on the other hand, was also an accomplished martial artist and possessed the rare combination of brain and brawn. Having pursued biology and medicine at London University, he also possessed a physique of Herculean proportions with the strength to match. He could have easily taken Marcellin apart, but instead remained calm and detached, watching the zoologist with mild amusement as he continued to make a fool of himself in front of Ebitas, Pelé, and the others.

Late into the night as I watched the embers of our camp fire slowly die, I thought about my three fellow explorers. Mark was 26 years old and loved the jungle. He spent all day on the lake and most of the evening catching insects in the forest. His in-depth knowledge on biology and medicine was amazing. At 20 years old, Joe was the youngest serving sergeant in the Territorial Army (the British equivalent to the US National Guard) and had certainly proved his worth on this expedition. Then there was Jonathan Walls, tall and well spoken, his intelligence and wisdom were well beyond his 26 years. He was a first-rate translator to boot. I was indeed fortunate to have such men as my companions on this dangerous adventure. I was the oldest of the group at 28 and by far the least qualified among us, and not at all worthy to call myself the expedition leader.

It seemed that our Congolese companions were relieved not to have seen or encountered a *mokele-mbembe*. Their reverence for this animal and the magical powers that some of them attributed to it was very evident indeed. At night, lowered voices muttered fearfully about invoking the wrath of the forest spirits. Those of us from the modern world of bustling cities, space shuttles, and cable television could laugh at such notions. But sitting by our campfire in the very heart of Africa, surrounded by the nocturnal melody of a million insects and other night creatures, only made the spiritual darkness of this primeval jungle all the more tangible. During our Congo adventure, I read the wonderfully entertaining book, *Eight Years with Congo Pygmies*, by Anne Eisner Putman, a New York artist who had lived with the little people in the Uturi Forest of the old Belgian Congo, not that far from where I sat. In her book, she described a terrifying encounter with an evil spirit known to the pygmies as the Esamba that dwelt in the heart of the forest, as she returned to her camp from a pygmy funeral.

> "Suddenly above the beat of the tom-toms I heard a clear mooing sound, half like that of a cow whose udder is too full, waiting to be milked, and half like that of a French horn, but pitched lower. I cowered against a mahogany tree, almost frozen with fear. It was the *Esamba*. I didn't have to be told. All the bits of idle talk, all the wild rumors, all the hushed whisperings of the natives suddenly became pieced together in my mind. Where the *Esamba* roamed, there went death. Its eerie, lowing voice was the voice of evil. At home I would have rationalized it and found some explanation to satisfy myself. Out in the rain forest it was something else again. Fear held me, vice tight. The false bravery that sustains one in as crowd drained away. I was alone, out of touch with the world I knew, a woman alone at night without even a jack-knife or hat pin to use against whatever foe might face me. I pressed my back against the bole of the huge mahogany tree, my heart pounding like a rock crusher. I was afraid to seek refuge in the pygmy village; just as afraid to make my way through the dark forest to the hotel compound. The mooing started up again. This time it was far to the right of the pygmy camp. How had it moved so fast and so far?
>
> "My tongue was dry. I couldn't swallow. Again the blood-curdling lowing came out of the darkness. It was farther around again. I thought to myself, 'My God, it's circling around to get between me and home.' Terror stricken now, I ran towards the camp, tripping over roots, catching myself on low-hanging liana vines. Shadows looked like roots and I dodged them. Roots looked like shadows and I fell over them. Branches tore at me like grasping hands. The voice of the

Esamba came from one side now. It was no longer behind me. I tried to run faster but I was no athlete. For years before I was married a few games of tennis had been enough to do me in for the day. Since then I had smoked far too many cigarettes to have any wind. Sweat oozed from my pores. I could feel my shirt, wet with perspiration, plastered tight to my body.

"If only I had my hurricane lantern. But like a fool I had left it at home, expecting some of the pygmy men to escort me back after the dance. I was crying by now, partly from fear, partly from the ache in my breast caused by the laboring of my heart. Subconsciously I asked myself why I was running for my life. All I knew of the *Esamba* I had gleaned in a dozen conversations around the campfire or from the natives whispered hints. Why was I, educated at the best of private schools, and a product of the Art Students League, racing through the Congo jungle to get away from nothing more than an unearthly, blood-curdling sound? Whatever the reason, I ran the way a man runs when he goes amok under the brain baking sun of the Indies. I stumbled, I fell, I gasped and I wept, but I kept on running through the jungle. Each time I heard the mooing it was closer. I knew I couldn't make it to the hotel compound. My only hope rested in the chance that I could last until I reached the village of the Africans, this side of the clearing.

"Once I stopped to listen and to catch my breath. My heart beat against the cage of my ribs like a wounded quail trapped in a briar patch. Then I heard the other noise. Off the path there was a rustling sound, as if an animal—or a man—were moving there in the darkness. Every vestige of civilization drained out of me and I fled screaming down the path, into the village and into the House of Abazinga, the animal keeper.

"He reached for his spear and sprang to the door, but I called him back.

"'No, no, don't go,' I cried. 'The *Esamba* is out there.'

"He turned to me, fear on his black face and in his eyes. 'Did Madami look upon the *Esamba?*' he asked. 'I heard it but saw nothing,' I answered, still fighting for breath.

"'Luck walks with you Madami,' he said. 'Anyone who beholds the *Esamba* dies within two days.' I sat by Azabinga's fire until I could walk again and he told me of the evil *Esamba*. Trees die when it passes, he said, and small bodies of water dry up. When the *Esamba* is loose, he swore, men hide their faces in their beds, lest they look upon its features and die. Seven nights the pygmies danced and cried for Basalinda. On the seventh night the little people held a feast. It was a ceremony to bring peace to the little body in the grave and to make her spirit content. The hollow wooden instruments rattled resonantly And the tom-toms beat long into the night. But the *Esamba* was silent."

The trek back to Boha took a mere three days. The brisk pace caused Jonathan and me to lag behind, our energy levels completely sapped by amebic dysentery. This did not sit well at all with Marcellin, who continued with his Dr. Jekyll and Mr. Hyde act. He would be amiable one minute and shrieking with rage the next. The longer we stayed in the forest, the worse his behavior became. During one brief rest in the forest, Marcellin told our guides that we were no better than the previous French colonials, out to exploit them, racist Britons who came to find *mokele-mbembe*, and make money on the backs of the poor, downtrodden Africans. Without warning, Ebitas, Pelé, and the others ran screaming towards us brandishing their machetes and spears, echoing Marcellin's words and accusing us of plotting to commit dreadful injustices toward them. After the initial shock of this unexpected and potentially disastrous situation wore off, we drew our machetes and faced them down. Jonathan verbally challenged them, speaking rapid French. Did we not pay them well for their help? Did we not show the proper respect to their chief? Did we not bring good gifts for their women and children? Did we not show the village elders the respect that they expected? Did we not obey all their laws? The anger and hostility of our guides quickly subsided, as Mark and Jonathan pointed out to them that as Marcellin's behavior was becoming more disturbing by the day, could anyone trust the words that he spoke?

Just two hours before sundown, we reached the village, thankful that we would not have to spend another night in the forest with Marcellin. Ebitas arranged our group in single file, in order of rank according to tribal etiquette and strict social order that everyone was expected to observe. Ebitas and Pelé would lead the group, with Marcellin next, then Jose with the rest of us behind. This arrangement did not sit well with Marcellin, who insisted on leading the group into Boha as he was the great scientist from Brazzaville. Later, the *Comite du Village* summoned Marcellin to explain himself and slapped him with a 5,000cfa ($10.00) fine for violation of village laws. Of course, the Prima Donna (as we began to call Marcellin) remained indignant at being treated in this way by common villagers and promptly demanded that we, the visiting *mondelis*, pay the money on his behalf.

After settling back into our brick home, Marcellin tried to intimidate us by standing outside his own quarters across the village square from us, thumbs hooked into his belt, glaring in our direction. Jose did the same, obligated to make a court jester of himself by mimicking his master. The scene looked like a movie take right out of *Gunfight at the OK Corral*, with the Prima Donna playing the role of the gunslinger.

After we managed to ignore him without collapsing into howls of laughter, Marcellin summoned us to a "meeting," in which he threw a tantrum and accused us of being "ignorant" of him, of Africa, of *mokele-mbembe*, and of everything else under the sun. He referred to us as "English children," and wanted to know the "real" reason why we had come to the Congo (which certainly wasn't to put up with his outrageous behavior). He then threatened to have us thrown in jail as spies, and wanted nothing further to do with us. Before bringing his tirade to a close, he warned us to leave the village women alone, as fornication was simply not tolerated here—especially as his reputation in the area was above reproach.

Two hours later as we sat in our quarters catching up with our notes over hot coffee, raised voices and a scuffle between several villagers drew us outside. One man was brandishing a shotgun and shouting something incomprehensible at Jose, while Marcellin and three other villagers attempted to restrain the gunman. After the situation had calmed somewhat, we had learned that Jose had engaged in carnal relations with the man's wife, thus provoking her husband into seeking out the Brazzaville buffoon with his shotgun. Delighted at this turn of events, we celebrated in style with hot soup and bread at Jose's inability to keep his pants on, while embarrassing the Prima Donna (whose delusional reputation was left in tatters). The offended husband accepted a hefty financial compensation for Jose's misdeed, and we would have gladly paid it just for the priceless entertainment it gave us. At least the incident happened at the end of our adventure. All we wanted now was to get back to civilization. The day prior to our departure from Boha, the chief and the elders summoned to his home, where, to my complete astonishment, I was presented with a tribal spear and pronounced "president" of Lake Tele to polite applause. The second surprise came when the chief asked me to make a speech. I stammered my way through a brief address, thanking them for their kind hospitality, how my fellow Britons and I had learned much from them, and that we hoped to visit them all again one day. As a bonus, Ebitas took us to a small pool in the swamp about a half-mile from the village where *mokele-mbembes* were known to frequent from time to time. But the dry season was in full swing and the pool had become a puddle. The honor bestowed upon me left the Prima Donna literally frothing at the mouth. How dare they honor a white man? A white man! He barely spoke to us after we departed from Boha the following morning. As there was no gasoline for the outboard, we were forced to paddle against the current from Boha to Epena, a backbreaking journey that lasted for 17 hours, with little rest in between. Arriving in Epena after sundown, we had barely enough strength left to unload our two *pirogues* and collapse into the brick "hospitality" building before heating up the last of our powdered soup. As we rested, an American appeared out of the darkness, having heard of our return from Lake Tele. Rory Nugent, a journalist and adventurer, was on his way to Boha, and eventually, he hoped, to Lake Tele. We sat and talked about our own experience, advising him to be sensitive in his dealing with the Boha villagers, and to try and focus his efforts on the northern end of the lake where at least two *mokele-mbembes* were known to frequent. Nugent would later write his own entertaining book, *Drums Along the Congo*, claiming to have observed a possible *mokele-mbembe* in the lake from about 1,000 yards distant.

News came the following day that there was no flight available from Epena to Impfondo for at least a month. The only alternative would be to paddle to Bimbo, then on foot to a new road currently under construction through the *Vassiere* by a Brazilian company. We opted for the journey, which would take us another two days by river and on foot. By this time the Prima Donna and the buffoon had abandoned us completely, but Jose did shout his farewells to us as he headed downriver to Impfondo. "See you in Brazzaville, Meester Beel," he yelled as the canoe disappeared round a sharp bend.

After a day's journey by river to Bimbo, we rested for another 24 hours before hiring four strong young porters to assist us in carrying our remaining equipment for the final leg. Another 16-hour trek through the forest brought us to the road being built by the Brazilians. To our relief, a large government truck was there, and ferried Mark and Jonathan back to the mission station where Gene and Sandy warmly welcomed them back. Joe and I would return to Bimbo on foot to carry another 100 lbs. of equipment back to the road the next morning. Although Joe was also beginning to suffer from dysentery, he found a rather unique cure. Sitting in the shade of a hut in the village of Bokele, Joe had purchased three large bunches of bananas from a local plantation owner and began to devour these with gusto. After forcing himself to eat 27 bananas in a single sitting, his dysentery literally disappeared overnight! After weeks of feeling miserable and lethargic, I followed Joe's lead and stuffed myself with the sweetest, juiciest bananas I had ever tasted. Surprisingly, my own condition rapidly improved with more energy and a distinct reduction in stomach cramps and dysentery attacks. To this day, Joe still proudly holds the banana-eating record!

Back at the mission station, I stood in the bathroom of the guest house and looked at the gaunt, bearded, thin young man staring back at me in the mirror. I had lost 25 lbs. Sandy patiently treated all our infections and ailments. The luxury of sleeping in real beds with fresh sheets, a daily shower, and good food in our stomachs soon had us back to our old selves. As we waited another week for the Linga Congo flight to take us back to Brazzaville, we pondered why we did not find any tangible evidence for the existence or otherwise of *mokele-mbembe*. We had spoken to eyewitnesses, examined locations firsthand where the animals had been active, and found that the fear of the animal was very real indeed.

One week later the plane finally arrived, and an hour after that, we were back in our small concrete dwelling in Brazzaville, ready to make our report to the British High Commissioner, Clive Almond. Marcellin summoned us to his office at the *Parc Zoologique* just to tell us how he regretted ever getting involved with us, and how we British should not be in the Congo, in spite of the fact that he had invited us there. Barely able to hold our tempers, we left the Prima Donna's office to attend our next meeting with François Ntsiba, the Secretary General, in order to present our own report to him, and to lodge a complaint against the Marcellin's disturbing behavior. His answer was typical of anyone with any kind of authority in the Congo: If you don't like the way we run things, then get out! In truth, we couldn't wait to "get out" of his ramshackle, backward country and its disastrously complex, complicated, and thoroughly corrupt Marxist-oriented system, where every petty bureaucrat had shamelessly tried to rip us off. Agnagna's threat to have us incarcerated as spies was the last straw. We insisted that the overly polite High Commissioner actually do something about the situation, and Gene Thomas, who was in Brazzaville on mission business, introduced us to a Brazzaville high court judge, Auguste Ikoli, who filed a court order against Agnagna, ordering him to cease harassing us and hand over all expedition equipment in his possession. Agnagna then took refuge behind Ntsiba, who demanded that "in fairness" to Agnagna's contribution to the expedition, we at least contribute some of our equipment to him before leaving the country. Reluctantly, we drew up a list, which included two cameras, a tent, two waterproof containers for camera film storage, items of clothing, belts, a large backpack and other sundry items. These were reluctantly accepted by Agnagna, whom, thankfully, we never saw again. Missing from our supplies, however, were seven rolls of film that were used to document our stay at Lake Tele and the wildlife around this beautiful and tranquil area—a unique photographic record now lost forever.

With more money forthcoming from London, we prepared for the long train journey to Pointe Noire, where a ship would be waiting to take us all home, thanks to Elder Dempster. The journey was the most uncomfortable I have ever experienced.

Only fourth class seats were available, and these turned out to be mere wooden benches screwed to the floor. As the train rocked and shuddered its way south overnight, a violent thunderstorm enveloped us, flooding the car with an inch of water, thanks to heavy rain pouring in through the open windows. Joe gave up trying to get comfortable and climbed into the overhead luggage rack where he managed to get some sleep during the journey.

Tired and disheveled, we arrived at the port town at 7:00 a.m., and sought out some breakfast after taking possession of our luggage. The Assistant British High Commissioner, Roger Simpson, was based in Pointe Noire as a director for the oil company, Conoco. Unwilling to allow us to fend for ourselves, Roger had ordered his staff to billet us in a company bungalow complete with servants and a refrigerator full of wonderful food and drink. For a week we were chauffeured around town, taken out for lunch and dinner, and entertained at the home of a wonderful German couple, Karsten and Klaus, who worked for the company, while waiting for our ship to arrive at Pointe Noire.

A week later our ship had literally come in. The 15,000-ton merchant vessel *Memnon* had docked with four air-conditioned cabins reserved for us on the return trip to London. As we gratefully extended our thanks and said our good-byes to Roger, Karsten, and Klaus, we boarded our ship and were led to our cabins by a steward from Sierra Leone. African crews had replaced British crewmen on almost all English merchant marine vessels. They were cheap labor to boot, and only the officers were British.

For four weeks we scrubbed decks, painted lifeboats, and performed other chores around the ship to kill time. Eventually we reached Tilbury, England, on a pleasant May evening. Joe, Johnny, and Mark had family waiting to greet them upon our return. I stayed with Mark and his family for two days before moving back to my old digs in London, and to my old job with the security company. As I walked along the familiar road with the 100 year-old Victorian terraced houses, watching children play soccer in the street, seeing familiar strangers go about their daily business, it all seemed a million light years away from my great jungle adventure. Would I ever return to that mysterious place?

The magazine *Fortean Times*, a publication devoted to all kinds of weird and wonderful phenomena, published a full account of our expedition. As I sat in the front room of the home of Bob Rickard, the Editor-in-chief, he showed me a photocopy of a letter to Richard Greenwell from Marcellin Agnagna, in which the Prima Donna himself openly admitted to deliberately leading us away from specific places in the river and swamps where we could have readily observed a *mokele-mbembe*. Agnagna also named some of the locations, but the details were blacked out by

Greenwell. So that was it. Agnagna knew perfectly well where perhaps even more than one *mokele-mbembe* could have been observed and filmed, but he chose not to share this information with us. We had paid him for nothing.

After my return from Africa I met the British actor Nabil Shaban, who had played a monster in the popular BBC TV series, *Dr. Who*. Nabil was keenly interested in Loch Ness and we later visited Tim Dinsdale upon invitation to his home in Tilehurst, Reading. We sat in the wooden shed behind his home, surrounded by all kinds of memorabilia from his decades of research at Loch Ness. Over hot coffee and cheese on toast, we talked about my Congo adventure. Of particular interest to him was my coming to Christ. As I recounted my experience to him and explained the enormous difference it had made in my life, I could see that my testimony had made a big impression on him. As a very young man, Tim was a follower of Christ, but life and all its trials and triumphs had led him on a path away from the cross. Two decades later as I write these words, I remember sending Tim a Christmas card in December 1987, only to find out that he had died from a massive heart attack merely days later. A true and inspirational friend was gone. Who would now carry on in his place at Loch Ness?

Also in 1987, Marc Miller, an anesthetist from Lancaster, Ohio, led his own expedition to Lake Tele. As I later learned from corresponding with Gene and Sandy Thomas, a dispute broke out over money with his Boha guides. Miller's refusal to pay more than the agreed price provoked his guides to abandon the Americans at the lake in the dead of night. Fortunately, a young Boha resident quietly made his way to the lake and secretly led Miller and his team to safety. They could have rotted there!

In 1989, the Japanese tried their luck and made their way to Lake Tele. The leader of the expedition later recalled how he and his team were literally held hostage by the Boha villagers until they stumped up more money than they had originally agreed to pay. Once again, *mokele-mbembe* proved elusive. Or were they, like our own expedition in 1986, deliberately led away from likely encounter spots by the inhabitants of Boha?

Also in 1989, one brief report from Gene Thomas rekindled my interest in returning to Africa. Gene and a team from the Evangelical Church in Impfondo to the village of Bendeko on the far northern reaches of the Likouala-aux-Herbes River. On the morning of December 30, the team quietly left the village to head south to Impfondo. As they slowly made their way through a large swamp pool, the team was startled when a large animal suddenly submerged in the center of the pool, sending foot high waves crashing against the mission vessel. The Congolese knew immediately that this was no hippo or elephant. Elephants cannot submerge completely and hippos are not found anywhere in that area. Unfortunately the heavy morning mist that blanketed the area ruled out any visual identification of the animal. Excited by this sudden encounter, Gene killed the outboard engine and waited for the animal to resurface. His Congolese companions, however, were in no mood to stay, as they were convinced that the animal would start looking for them, or perhaps even surface under their canoe and ditch the entire team into the water. The prospect of encountering sizeable crocodiles inhabiting the swamps and rivers—never mind a ferocious *mokele-mbembe*—was simply not on, and Gene reluctantly abandoned his vigil.

In 1990, the Congolese government grew tired of being bombarded with complaints from the various embassies and diplomatic missions that represented foreign nationals ripped off by the Boha villagers. To bring the matter to an end, President Denis Sassou-Nguesso dispatched his soldiers to arrest the chief of Boha. He was not running an independent state and would be brought to heel once and for all. If I was to return to the Congo, it might be better to avoid Boha for the time being. Yes, I had made many friends there, but if the situation remained volatile it would be best to continue my research elsewhere.

It was now winter of 1991. Six years had passed since my return to Africa. I had met my wife Terri, a student nurse of Chinese extraction from Mauritius, in November 1986. We were engaged in the summer of 1987 and married in July 1988. Our handsome son, Matthew, was born in October 1991, and was thoroughly spoiled by all his relatives. As the years progressed, I corresponded frequently with Gene and Sandy, who had been relegated to Brazzaville by the United World Mission to develop a city church. Paul and Diane Ohlin from New England had replaced them as the principle missionaries in Impfondo. Roy Mackal and Herman Regusters had abandoned their plans to return to the Congo with new attempts to find *mokele-mbembe*, and Redmond O'Hanlon, a well-known British adventurer, had written an account of his own trip to Lake Tele with Marcellin Agnagna in search of *mokele-mbembe*. In his book, *No Mercy: A Journey to the Heart of the Congo*, O'Hanlon recorded his own observations about Agnagna's odd behavior, and stated his belief that *mokele-mbembe* probably did not exist and was most likely a case where local hunters and fishermen misidentified elephants crossing the rivers and lakes with their trunks held in the air. This was the very best way to insult a pygmy. The pygmy and Bantu tribes of Equatorial Africa were hunting elephants long before the arrival of the white man and knew perfectly well what a swimming elephant looked like.

There was a vast difference, of course, between a swimming pachyderm and an elephant sized, reddish-brown semi-aquatic monster with a long thin neck ending in a distinctive snake-like head topped with a frill like a rooster. Elephants do not possess long, thin tails, nor can they submerge under swamps and rivers for long periods of time, only to occasionally surface under canoes, spilling the occupants into the water and killing them by tail lashing and biting. O'Hanlon and others have every right to be skeptical, but just because they did not observe a *mokele-mbembe* for themselves does not mean that the animal is a misidentified elephant or just a creature of mere imagination. 100 years of evolutionary brainwashing had done its job well. Dinosaurs, we are told, died out about 65 million years ago and the genus *Homo* did not appear on the scene until about two million years ago. Therefore, no human being has ever seen a living dinosaur. Some dissenting voices in science were not so quick to accept this logic, but Sanderson, Mackal, Heuvelmans, and other gadflies were in the minority.

1992

I had made up my mind to return to Africa, but this time I would combine a humanitarian mission with the hunt for *mokele-mbembe*. Gene and Sandy had become frustrated with the theft of medical supplies and other items that were sent from the USA to the Congo for use during their field trips. Almost everything would be stolen as soon as the shipment arrived in Brazzaville. Could I help? The biggest stumbling block was, of course, money. I needed at least five thousand pounds sterling ($12,000 US) to conduct even a modest trip to the Congo. As before in 1985, news spread quickly of my return to Africa, attracting the attention of the media. Newspaper reporters were appearing on my doorstep, and one national newspaper, *The Daily Star*, dispatched a professional photographer to Kent where I donned my jungle clothing, hung a pair of binoculars around my neck and tried to look convincingly adventurous during a photo shoot in a local park. The exposure attracted the attention of Candy Thorpe, who ran a creative writing company in London called Lexicon. I was becoming somewhat embarrassed by all the attention, as I had not even managed to raise 100 pounds sterling towards the proposed second expedition. Candy explained that she was acquainted with the business manager of Bill Wyman, the guitarist from the Rolling Stones Rock Group, and suggested I contact him regarding my efforts to raise funds for the trip. I was taken aback by her suggestion. Bill Wyman? Why would he ever want to give money to help out missionaries? At first I just could not even entertain this idea, but appeals to local churches and an interview on a popular local Christian radio show did little to fill my expedition coffers. With great reluctance, I wrote a letter to Bill Wyman, explaining the thefts of medical supplies and other much needed items and equipment. I included a list of all items I needed to purchase, including photocopies of letters from Sandy Thomas, which spoke of "taking the glorious Gospel of Christ to the heathen tribes living in the bondage of Satan and sin."

"Yes, Bill Wyman is going to love this!" I thought to myself as I mailed the envelope to him at his recording studio in London. Not expecting much, if anything in return, I was astonished to receive a letter from Wyman two weeks later, expressing an interest in the medical work currently being conducted in the Congo. And yes, he would be happy to help with funding! I almost fainted with shock. Here was a guitarist from the world's most famous rock band offering to help with funding for missionary work! I don't believe that Bill Wyman was particularly religious, but what a wonderful gesture. Before parting with any money, however, Bill wanted to see some character references, just to be certain that I was on the level with him. I quickly conferred with my pastor, Chris Voke of Gillingham Baptist Church, who provided me with a written reference and commendation for the Congo project. This in turn was mailed to Bill Wyman for his approval. Two weeks later, a large envelope arrived containing a number of checks from the legendary rocker and other famous musicians and personalities. These individuals and the amounts they donated (in US dollars) were:

Bill Wyman	$2,000
Mick Jagger	$2,000
Ken Follett (novelist)	$2,000
Charlie Watts	$1,000
Ringo Starr	$1,000

Not only was the $8,000 enough to send me to the Congo for at least a month, but two other organizations came through with generous donations of medical supplies and educational items for the mission clinic and school in Impfondo. ECHO, based in London, supplied four large cases packed with medical supplies, and the Bible and Tract Society donated 100 New Testaments in French, plus 1,000 individual Gospel booklets and tracts in the Lingala language. These things were truly an answer to prayer. A small shipping company in London was able to provide inexpensive air cargo for our supplies to Brazzaville via Air Portugal, and I secured a return flight from London to Brazzaville via the Russian airline, Aeroflot, for $1,000.00. I would depart for the Congo in October, and my wife would fly to Mauritius for six weeks with our infant son, who would no doubt be the center of attention among all his relatives there. As I sat at home on a dreary Saturday morning, gazing out of the window at the scarlet leaves being tossed around by the gray gusts of the autumn wind, my heart missed a beat as I thought about the adventure ahead. Africa! Now there was the place to be!

Once again a flurry of media attention led to more radio and newspaper interviews, including a feature on my return to the Congo by BBC Radio 4, which was broadcast all over the country. Letters began to arrive from would-be explorers and adventurers from all over Britain, asking how they might join the dinosaur hunt in Africa. I was becoming rather nervous at this point, as the bulk of the expedition funding had come from Bill Wyman, who wanted to see some humanitarian good come out of the trip. He would not be disappointed.

Among the many letters that had reached me was one from Liz Addy, an optometrist from Ilkley, Yorkshire, who had traveled all over the world engaging in humanitarian work, treating eye infections, and providing free eyeglasses. I called Liz and after a short telephone interview, we agreed to meet in London at the British Museum of Natural History a few weeks later. Liz turned out to be nothing like I had imagined her. Small and slim, her eyeglasses and shoulder-length mousy-blonde hair made her look like a typical librarian. Given her love of paragliding, mountain climbing and world-wide travel, she would be an adventurous bookworm to be sure! Liz was keen to visit Equatorial Africa and put her optometric skills to good use. I assured her that she would have little rest, given the great need for all kinds of medical work in the Congo. We agreed to move forward and Liz returned to Yorkshire where she sent me a check for $1,000.00 to cover her airfare. With our tickets booked, I visited Mr. Louis Muzzu of the new Congolese diplomatic mission in London, where our visas were quickly secured without the nonsense I had to endure in Paris seven years before.

Gene and Sandy were still on furlough in the USA and would arrive in the Congo a few weeks behind us. Our main contact was another American, Lt. Colonel William Collins of the Salvation Army. He had arranged to pick us up from the airport and billet us at the Swedish Mission, a beautiful colonial period building constructed in 1902, which still served as the administrative center for Swedish missionaries serving in the Congo and other locations in Equatorial Africa. This time we would be working strictly with the missionaries. There would be no corrupt government ministries to deal with and no outrageously long waits for paperwork or ridiculous demands for inflated "fees." But Africa being what it is, even the most carefully laid out plans could still go badly wrong. How would we fare?

At last we were ready to leave for Africa. My inoculations were updated and malaria pills packed into my first aid kit. With one backpack and one suitcase packed and ready to go, I caught the train to London and met up with Liz. Her brother, Brian, would put us up for the night at his home in Middlesex and drive us to

Heathrow airport the following morning. As we relaxed in Brian's home, BBC Television broadcast disturbing pictures of an Aeroflot airliner—a Tupolev 154—that had crashed shortly after take off somewhere in Russia due to overheated engines. The pilot attempted to return to the airport but crash-landed in a snow-covered field. The plane hit a farm building and exploded, killing everyone onboard. I didn't get much sleep that night, knowing that we were about to fly on Aeroflot. The trip to the airport and booking in our luggage, bound for Africa, went smoothly. As we waited in the departure lounge, I was surprised to see that the Aeroflot plane we were about to board was an Airbus A320. Aeroflot had decided to upgrade its fleet by purchasing Boeing and Airbus aircraft. The mighty USSR had imploded and shrunk. The Soviets were no longer able to afford the independent development of new airliners. Only the Russian military industry was building new aircraft for their air force. The Cold War might have been over, but the Russian bear still had sharp claws.

The flight to Moscow was pleasant and without any dramatic incidents. Moscow International Airport, however, was the most depressing place I had ever been. Half the airport was without electricity, the public bathrooms were absolutely filthy, and one floor of the airport seemed to have been taken up by a thronging mass of refugees from Somalia, desperately seeking refugee status in Europe. For some odd reason, the airport security staff had detained the refugees, rather than let them continue to Western Europe. Desperate to find some refuge within the depressing mausoleum that masqueraded for an airport, we found, of all things, an Irish restaurant called Mullins. Here we were able to purchase large plates of delicious Irish stew and a pot of tea with our British pounds. This little bit of Irish heaven was run by a couple from Dublin who lived in Moscow. After a few years they would return home and be replaced by another Irish couple. To kill the boredom, I engaged some of the airport employees in conversation. Life was tough for them, especially the young who wanted to escape from their dreary lives and move to the West where there were good opportunities waiting for them. How fortunate we were never to have suffered under communism.

After our 12-hour stopover, we were grateful to be boarding our flight to Brazzaville. Looking out at the rows of snow-covered Russian airliners, I wondered how many of them were actually airworthy. As we shuffled along in line, presenting our boarding passes and passports to a dour-looking Russian security officer, I glanced out of the departure lounge window to see what kind of plane we would be completing the last leg of our journey in. It was a Tupolev 154—the same type that had crashed only a few days before. I mentioned this to Liz, but she remained philosophical and just wanted to get to Africa. Like most Russian built planes, the Tupolev 154 was a virtual copy of a Western aircraft design. In the case of the Tupolev, it was identical to the Boeing 727, a medium-range airliner with three engines located at the back of the aircraft. How well the Russian design would compare to its American counterpart would soon be seen. As we sat looking out at the bleak Russian winter, the plane taxied out onto the runway. After a short stop, we were hurtling down the runway, the distinctive whine of the Kutnetsov engines screaming at full power as we left the ground 40 seconds from a standing start. Judging by our rate of climb, I wondered if the captain of our airliner was a former MiG fighter pilot set on intercepting waves of imaginary enemy bombers invading Russian airspace.

As we reached our cruising speed of 500 mph at 37,000 feet, I settled into my seat and took stock of the interior of this workhorse of the Aeroflot fleet. The interior was sparse. There was no in-flight movie, music, or entertainment of any kind. The food was bland and the toilets smelled revolting. Passengers were expected to provide their own entertainment. I listened to Enya on my Walkman, while Liz slept. Some passengers read books while others played cards. The African passengers sang, argued, and snored loudly.

Eight hours later we landed in Bangui, the capital of the Central African Republic. Disembarking from the plane and into the cloying heat, the sights and sounds around me flooded my mind with so many bittersweet memories from 1986. Liz committed the cardinal sin of whipping out her video camera and filming the plane as it was being refueled. Within a few seconds we were surrounded by African Gendarmes shouting at us in bad French and grabbing Liz's camera. We were filming in a civilian airport, which was strictly forbidden. Rather than allow ourselves to be arrested as spies, Liz handed over the film cassette from her camera and demanded a receipt. The only receipt African police officers give are in the form of handcuffs and beatings. After the situation had calmed somewhat we were allowed to board the plane and continue on our way.

One hour later we touched down at Maya Maya International Airport in Brazzaville. In spite of the shortcomings of Aeroflot's in-flight service, the Russian pilots proved to be superb aviators by treating us to the smoothest landings and take-offs I had ever experienced with any major airline. Somehow it was almost joyous to return to the Congo. After passing through the passport check, we looked around for Colonel Collins, only to find that no one had arrived to meet us. We met an immaculately attired Congolese pilot in a small office, who kindly allowed us to telephone the Salvation Army headquarters and arrange for someone to pick us up. 30 minutes later Colonel Collins himself showed up with a jeep and drove us to the Swedish Mission.

The drive through Brazzaville filled my mind with so many memories. The sights, sounds, and smells of Africa filled me with excitement. Perhaps this time we would make the breakthrough in the search for *mokele-mbembe*. Liz was excited to be in Africa for a different reason. This was a new experience for her and a chance to expand on her work in another part of the world.

The Swedish Mission was a sight for sore eyes. A white colonial oasis in a bustling city, there were several Swedish missionaries in residence, enjoying some rest and recuperation from their months working in the former Belgian Congo (Zaire) across the river. The curious practice of the Swedes that involved eating their main meal (dinner) at mid-day, followed by a light supper in the evening, took a bit of getting used to. But Liz wanted to see Brazzaville, and after dumping our bags in our rooms, we borrowed some CFA from a resident missionary and walked into town. Sitting in a French bar, we enjoyed cold drinks as the evening rays of the setting sun cast a golden glow over the mighty Congo River as the local fishermen slowly paddled their canoes laden with fish back to the bank.

The following morning we received word that our cargo had not yet arrived in Brazzaville. To make the most of the day, Liz and I again decided to visit the city and take in the sights. During breakfast I casually mentioned that I would like to visit the Brazzaville Catholic Cathedral, a comment that provoked wide-eyed stares among the missionaries at the table. A Catholic place of worship, they wondered? I quickly explained that I would not be actually worshipping there, but merely photographing the unique 1950s French architecture. With sighs of relief, the Swedes continued with their breakfast without further ado. I wondered why they were so shocked that I would visit a Roman Catholic place of worship, considering that they spent their evenings watching James Bond movies, full of violence, killing, and scantily clad glamorous "Bond girls."

After a week at the Swedish Mission and still no sign of our two crates, I began to have nightmarish visions of 1986, when we spent months waiting to get our equipment and gear before we could even being to prepare to head into the bush. A quick trip to the offices of Air Portugal determined that our supplies been delayed due to a high volume of cargo bound for Brazzaville, but we would have our cargo within the next few days. To save money, we moved from the Swedish Mission to the home of Mac and Marnie Wigfield, Canadian missionaries who were working in Brazzaville. They rented a spacious home from a Congolese Army Colonel and kindly allocated two rooms for us.

Finally, with the arrival of the cargo, we were off on a flight to Impfondo. Liz, who could never sit still for very long, struck up a conversation with the French pilot, who allowed her to film in the flight deck. By mid-day we began our descent towards Impfondo, as our plane rocked and shuddered through pockets of low-level turbulence in an effort to reach the ground. We wheeled low over the Ubangui River, and my heart gave a leap as we landed on thin ribbon of blacktop. At 10:45 a.m., we had finally arrived in Impfondo, that special place where I had met with the Creator of the universe six years before.

We waited patiently as our baggage was retrieved from the cargo hold, and an old friend was there to meet us: Pastor Mowawa Eugene, who had baptized me in the Ubangui River six years before with Gene Thomas. As we hugged and greeted each other, a familiar voice from the past called my name from within the midst of the thronging crowd. "Meester Beel! Meester Beel!" Scarcely wanting to believe my ears, I turned to see Jose Bourges, the lovable Brazzaville Buffoon, grinning from ear to ear as he pushed his way through the crowd towards me. "What are you doing here?" he asked. "Oh, just delivering a few supplies to the mission station," I replied, trying to make it sound like a visit to the supermarket. Jose had been promoted to Chief Game Officer for the Epena Region and was now located in Impfondo. Marcellin Agnagna was thankfully still in Brazzaville and unaware of my return to the Congo. A customs official approached our group and quickly relieved Liz and me of our passports, instructing us to report to the *Chef de la Region* the following morning. Paul Ohlin, the missionary who had replaced Gene Thomas in Impfondo, arrived in a Land Rover and ferried us to the mission where we shared a house with Sarah Speer, a missionary nurse from Winnipeg. After dumping our baggage in our new quarters, we headed for the main mission house where Diane Ohlin had kindly prepared lunch for us. As we tucked into fish and chips served in honor of their British guests, we discussed the Ohlin's desire to overcome some formidable barriers in order to establish a culturally appropriate church among the Aka people, the local pygmy tribe.

Ever keen to explore a new place, I showed Liz around the town of Impfondo before heading back to the mission station before sunset. We visited the pygmies that lived in the forest on the edge of the mission station with Jose, Paul and Diane's five-year-old daughter. For such a young child, her insight into the culture and her command of Lingala was truly amazing, to say the least. As I settled down in my room for the night, it didn't seem as though it had been six years since I last resided in the mission guest house. Tomorrow we would meet with the local *Chef de la Region* and arrange our trip into the interior.

Our final day in Impfondo went uneventfully, as we met with the *Chef de la Region,* a friendly man somewhere in his thirties who spoke very good English. After examining our passports and, he asked us a few basic questions regarding our visit to Impfondo before deciding that we were harmless enough to accompany Sarah Speer on a trip into the interior. The remainder of our day was spent purchasing food supplies and preparing our backpacks for an early departure. The sun began to

Tom Hall, Sarah Speer, Pastor Matena Paul, Liz Addy, Bill Gibbons

Swamp Scene

dip beneath the jungle wall, and I waged a losing battle against the dwindling light as I hurriedly stuffed my backpack with spare clothing, emergency rations, cameras, spare film, batteries, binoculars, a first aid kit, and essential items, including the all-important roll of toilet tissue. On the morning of October 16th, we were all up before dawn and making our preparations to leave by the light of our gasoline lanterns. Less than an hour later the emerging sun had given us just enough light to take a group photo around the mission Land Rover before the long drive to Epena. As we sped along the new tarmac road to Epena, I thought back to the exhausting, multiple treks that I made in 1986, carrying our expedition supplies along the narrow forest trail that wound its way through jungle and swamp to the dirt track where we caught rides from the Brazilian road crew and the Congolese government truck.

We arrived at Epena before 8:00 a.m. The town was already a hive of activity as Sarah and Liz made their way over to new Catholic mission, where two bearded French priests were dispensing medication to a small population of townspeople and nearby villagers who suffered from a variety of maladies, some quite serious. A young boy showed me a tiny python that he kept in a glass jar partially filled with water.

When I explained that the snake would one day grow big enough to swallow him whole, he laughed and skipped happily away to join his friends. Making my way over to the Catholic mission, Sarah announced the happy news that the priests had agreed to loan us their aluminum riverboat for our trip. At fifteen feet in length, it was a sturdy craft, well balanced with watertight storage areas fore and aft with sturdy benches to sit on. What a difference from the leaky, precarious, narrow wooden *pirogues* that we had to endure on the previous expedition!

Thankfully, there was sufficient gasoline to purchase for our outboard engine, kindly supplied by Paul Ohlin. An old battered tin pan was used to measure out the gasoline as Pastor Matena Paul carefully filled our large Gerry cans, spilling nary a drop. Soon our boat was loaded with our backpacks and supplies, ready to go. With a sharp pull of the cord, our engine roared into life and we were off.

Excited village children ran along the riverbank, waving to us as we waved back, slowly leaving them behind. The end of the rainy season had swollen the rivers to the point that much of the land had been waterlogged, making conventional navigation by map almost impossible. Although the pastor knew his territory well, we had to seek directions on at least two occasions from Congolese travelers who were also making their way through the flooded plain. However, several hours of travel time had been cut from our journey as our marvelous aluminum boat sliced effortlessly through the flooded landscape with ease. We quickly passed familiar villages along the way, including Itanga and Boha. Although I had made many friends in Boha, we all thought it would be wise to avoid landing there for the time being, as the village was a flashpoint for trouble in the past few years. Moods could change quickly and any friendly visit from white outsiders could be misconstrued and become dangerously confrontational if tribal and cultural protocol were not carefully observed. Besides, word had reached us of a Japanese expedition that had recently visited Lake Tele. They were taken there after negotiating with the elders of Boha village, who promptly held the expedition hostage while the sum of $12,000 (USD) was sent from Brazzaville to secure their release. Yes, we agreed, Boha was best avoided for now.

We continued and eventually landed at Dzeke, a sizeable village and the home of the hunter and *mokele-mbembe* witness, Emmanuel Moungoumela. After making our way through the usual crowd of curious children to a hut reserved for guests and dumping our supplies in a small bedroom, I lost no time in making enquiries about Mongoumela. Was he here? Would I be able to visit with him? A young Congolese man who spoke excellent English approached me and introduced himself as the village school teacher. His job was to teach French to the children in an area where Lingala was most commonly spoken. He was delighted to see westerners visit his village and said he would be happy to translate any information that Moungoumela may have for me.

Later that afternoon, Emmanuel Mongoumela appeared at our hut. He had never fully recovered from his previous illness and had retired from elephant hunting. He spent most of his time fishing on the river to make what money he could to help support his family. Had he seen *mokele-mbembe* since my last visit? As we spoke, two village elders appeared and were intent on listening in on the conversation.

No, he had not actually seen *mokele-mbembe* but had found traces where the animals had been. Over the past six years, he had observed claw marks on the riverbank where the molombo fruits had been stripped away from the vines, great breeches in the foliage where the larger specimens had pushed their way through to reach their food supply, and clawed footprints deeply imprinted in sand banks where the animals had crossed moving from shallow to deeper water. Although he was certain that no other animal devours the fruits, I still asked him if the footprints could the footprints still have been made by an elephant (*nook*) or a hippo (*nub*) in the area.

"Of course not, you foolish white man!" he retorted. He knew perfectly well what elephant prints looked like, and hippos are never found in *mokele-mbembe* territory, especially around the confluences of the Ndoki and Sangha Rivers, or the Assomba and Sangha Rivers. Occasionally, he would detect the movement of some huge animal moving around in these specific locations, shrouded and hidden by the early morning mist that surrounded the Likouala swamps just before sunrise. But the once fearless hunter no longer ventured into the more remote areas where he knew *mokele-mbembes* could be observed. His frail condition had perhaps brought him to a realization of his own mortality. He was no longer willing or able to help us locate and film a *mokele-mbembe* and incur the wrath of the *ndami*. However, another hunter mentioned that he and a friend from the village had spotted an *emela-ntouka* south of the village near the river on a few months before. Once again the description of this particular animal remained accurate in every detail. The large hippo-like body, the long, polished horn on its snout, the short tail, and its ability to submerge completely in the river. At least one of our mystery animals had been seen, and very recently. This ferocious beast was another reason why hippos avoided the area. The *emela-ntouka* attacked and killed hippos (*ngubu*), elephants (*nzoku*), buffalo, and even local fishermen if they ventured into a stretch of river where the horned terror lurked. Bringing our conversation to an end, perhaps due to the presence of the *ndami* (village elders) my old friend shuffled away into the dwindling light as the aroma of hot soup beckoned me back to my own modest dwelling for supper and a welcome bed. Tomorrow we would meet the *ndami*. Would they welcome their white visitors to their village?

Our morning began with a bowl of oatmeal and a mug of piping hot tea. A great way to start the day in the jungle! Sarah and Tom left to see if there were any visiting doctors or nurses in the village, while I sat near the door to catch up with my notes. Four members of the *Village du Committee* showed up, but as I was the only one there they would return later to meet with the rest of the team.

After a modest lunch of fruit, bread and coffee, Liz set up her eye clinic and was soon swamped with patients. Sarah, Tom, and Moise Molembe, a medical technician who had joined us at Dzeke, established a clinic and had the remaining village population in attendance. I wondered if there was anyone left in the village at all who did not have some form of ailment, real or imagined.

Both clinics were still open well into the late evening. Liz dealt with common eye ailments and dispensed dozens of free spectacles, while Sarah, Tom, and Moise labored away, diagnosing and treating everything from malaria to yaws. One young man hobbled painfully across the village with a bloated right leg. He was clearly suffering from a horrid case of elephantiasis, a disease that is characterized by the thickening of the skin and underlying tissues, especially in the legs and genitals. Elephantiasis generally results from the obstruction of the lymphatic vessels and is most commonly caused by a parasitic disease known as lymphatic filariasis. As there was little we could do to ease his dreadful suffering with our limited medical supplies, we encouraged the young man's family and friends to arrange his passage to

Brazzaville as soon as possible in order to receive urgent treatment for his condition.

Exhausted by hours of medical care they had invested in the inhabitants of Dzeke, our four dedicated pioneers literally flopped into bed as soon as our modest supper was over. I stayed awake a little longer in my small but comfortable bedroom, catching up on my notes and feeling somewhat useless, as I possessed no medical skills to offer to ease the burden of my dedicated colleagues. And yet, I wondered, what exactly would it take to provide medical care to the villagers that resided along the river system here? All that would be needed would be one medical doctor and two nurses based in Epena. They could conduct river trips to the villages, perhaps every three months, treating the usual maladies and illnesses. This would cost the government money, having to bother itself with the tedious task of providing the Likouala region with medical staff, a motorized boat and a sufficient cache of medical equipment and supplies. The cheaper option, of course, was to allow western missionaries to build and operate schools and clinics for free, without burdening the government to actually spend money on its own people.

The following morning Sarah and I attended a Sunday morning service at the village church. A typical church in a Congolese village was nothing more than a thatched roof on stilts. The congregation would sit on hard wooden "benches" made from tree trunks split in half, and everyone sang popular hymns translated in Lingala to the beat of a drum. After the singing, Sarah got up, introduced me to the assembly and promptly invited me to speak. Me? What would or could I say?

Then it came to me: Give them your testimony! And so, I began to speak, telling my story of salvation and finding Christ in the jungle, as Sarah translated everything into Lingala. After I had finished speaking, Sarah gave an altar call, and three people, an old Congolese woman and a young boy and girl, perhaps in their early teens, came forward and accepted Christ. We laid hands on them and joyfully prayed for them, knowing that the angels would be celebrating the redemption of three precious souls on this Sabbath day. The power of Satan had been broken. It was a glorious day indeed!

Later that afternoon, we had packed our belongings and loaded up our boat. The usual large crowd had gathered to bid us farewell, and one grateful villager even gave Liz a live chicken as a parting gift for all her work. Liz must surely be the first optometrist to have ever visited this area. With our parting hugs and good-byes finished, we headed for our next destination, the village of Kinami.

None of our party had ever visited Kinami before, and my only reference to the village was in Roy Mackal's book, recalling his encounter with friendly, but cautious people who did not seem at all willing to discuss or share their knowledge of *mokele-mbembe*. We found this to be very much the case when I attempted to draw some of the village fishermen into conversation on any strange or odd animals that they may have encountered at one time or another without referring directly to *mokele-mbembe*. Another reference was made to a recent encounter with the *emela-ntouka*, independently verifying Young Emmanuel's report from Dzeke.

Later during another marathon clinic with Liz, Sarah, Tom, and Moise hard at work, a young village teenager approached me in the darkness of the cool evening. He explained that *mokele-mbembe* was well known to the village elders and the fishermen who work on the river, but they do not discuss the fabled beast with outsiders. It had also been seen recently, and most often in the wet season, but hardly ever in the dry season. What was this relationship with the seasons, I wondered? The people were in agreement about one thing; *mokele-mbembe* was a monster that feared nothing, and was therefore to be feared above all other animals, whether known or unknown. It was the god of the river!

After 24 hours at Kinami, our medical work was completed. More assorted illnesses and maladies had been treated; more eye conditions and free spectacles had been dispensed. It was time to move on.

Tom had heard that there were herds of elephants in the swamps to the northwest around Lake Tebeke. A break to do some sightseeing north of Kinami was an interesting idea, and we decided to press on with Mauree, a Christian fisherman who had offered to be our guide to Lake Tebeke. The farther north we traveled, the more primitive and beautiful the scenery became. Feeling drowsy as the hot sun beat down upon us; I was suddenly startled by the sight of two colobus monkeys rapidly swimming across the river just ahead of our boat. It was surprising indeed that they had not been snapped up by a crocodile, as one had submerged just ahead of our boat seconds before the monkeys had decided to qualify for the Congolese Olympic swimming team.

As we progressed farther up the Bai River, Pastor Matena Paul pointed to an odd-looking line of trees to the east. "That is the way to Lac Tele, through the swamp," he stated. Five narrow water channels led from the Bai to the lake, but they were often blocked by fallen trees and partially submerged log jams. The only way to the lake was overland on foot from Boha. However, Lake Tele was easily accessible from the air if one could afford to hire a floatplane. This would greatly ease the journey to the lake and give any expedition much more time to fully explore this remote body of water.

Moise suddenly veered northwest on a small channel of water. Within minutes we were slicing through the calm waters of a small lake. "Lac Fouloukou," said Mauree, anticipating my question. "*Mokele-mbembe* has been seen here," he added, matter-of-factly. We continued on, urged by Tom who was keen to see wild forest elephants. No sooner had we exited the lake, a dense thicket of bush and fallen trees completely blocked our way. It seemed unlikely that we would be able to continue, but Tom and Liz began to hack their way through the dense bush with gusto, eventually creating a breech through the tangled mass. Slowly we passed through an area of dark, forbidding swamp until we reached dry land and a welcome clearing. This was a temporary camping area for fishermen and hunters, and an ideal place to pitch our tents for the night. The time had come for our chicken to serve as dinner. Liz, a vegetarian, would certainly not be willing to kill the bird. Sarah also declined the task, and reluctantly, so did I. Tom, always ready for anything it seemed, picked up the bird, grabbed his machete, and walked off into the forest to do the dark deed. Tom returned to the camp five minutes later with a chicken minus its head, and Celastin busied himself preparing the bird for dinner. I wondered what the others thought of me—the fearless Scot, the great explorer, out to find his dinosaur but too squeamish to kill a chicken.

By daybreak we headed into the swampy forest on foot. Celastin led the way, and casually stepped into the swamp up to his knees. We all followed suit, and waded slowly towards a small *pirogue* that was half-filled with water. After this was emptied, we climbed in and Pastor Matena Paul slowly punted us one by one into the swamp. Standing up to my knees in swamp, listening to the cacophony of sounds that echoed around me in the primeval beauty of the Congo, I wondered what strange, wonderful, weird discovery we might make?

Eventually, the pastor returned and ferried me in the narrow, wobbly canoe to another body of water, called Lake Tebeke. This was a smaller than Lake Fouloukou, but too big to be a swamp pool. The others had decided to explore the immediate forest area, something that Mauree was unhappy about, as we possessed no firearms. Nevertheless, the team headed off into the gloom while I decided to stay at the lake. Surrounded by a dense wall of vegetation, it was almost impossible to determine if any large animal was in the immediate vicinity. A nearby tree provided the answer as I shinnied up the trunk and carefully climbed onto a branch overlooking the lake. The entire body of water was surrounded by a wall of vegetation. Elephants would find it difficult to reach the lake, but a shy, semi-aquatic animal like a *mokele-mbembe* would find it an ideal habitat to feed and hide away with little or no chance of being disturbed by elephants, hippos, or humans. With all the grunting and heaving I had made in my struggle to find a vantage point to get a better view of the lake, any wary, elusive animal would have been alerted to my presence and made its retreat into the swamp. Surely this was where *mokele-mbembe* would be found? The lake was remote, tranquil and in the right area. Elephants, on the other hand, would more likely be found at Lake Fouloukou, which was closer to the Bai River and open savannah. Later, we would learn from the pygmies in Impfondo, including Marien Nkoli, who was from that area, that *mokele-mbembes* had indeed been observed in Lake Tebeke and the surrounding rivers and swamps. Celastin later mentioned that he had observed a gigantic 10-meter crocodile in the Bia River only seven months before. He was certain that this was not a mere outsized Nile crocodile (which he had observed countless times), but the monstrous *mahamba*.

After a chicken dinner and welcome night's sleep, we were ready to return to Impfondo. Our fuel for the outboard was running low, so we elected to take turns and paddle, which was surprisingly easy given the size and sturdy construction of our aluminum boat. The paddling turned out to be relatively easy as we were floating along in the same direction as the current. It also gave us a chance to sit back and enjoy our surroundings more and take in the amazing scenery. By 4:00 p.m., we decided to fire up our outboard and find a place to camp for the night. Just before the sun disappeared, we found a fishing camp and prepared to bed down for the night. Liz busied herself by examining the camp, and noticed that the two wooden huts appeared to be "moving." To the shock and dismay of the entire team, the camp had been overrun by millions of army ants. They were everywhere, slowly making their way through the camp and into the forest beyond. We could do little else but stand near the river and watch as the miniature invaders quickly consumed every spider and insect that had inhabited the huts. Two hours later, the ants had almost all gone, leaving behind huts picked clean of anything edible. A few stragglers remained, persuading us to pitch our tent outside and make a fire for cooking. No sooner had we settled down than another party arrived at the camp for the night. This was Celastin Makombe, the chief of Toukalaka, who was insulted to learn that we had visited Lake Tebeke without his permission. He demanded to know who we were and wanted to examine our papers. Pastor Matena Paul and Sarah engaged the chief and his entourage in conversation, keeping them busy with "talk," which is basically idle gossip that the Congolese love to engage in. Sarah went on to explain, with theatrical style, how our medical supplies were delayed getting into the country, how we had to wait patiently, how we had to pay extortionate rates to fly to Impfondo, and how all we wanted to do was to bring good medicine and spectacles to the people living in the Likouala. Our visit to the lake was merely sightseeing, and we could not visit the chief if he was not at his village, anyway. But we left him with a good impression and hope to visit with him again one day.

After a few hours of "talk," we all retired to our tents in good spirits, with a few stray ants to keep us company for the night. Liz and Sarah decided to rise before dawn and plunge into the river for a refreshing swim. Knowing that sizeable crocodiles have been observed in almost all Likouala rivers, Tom and I stayed on dry land and gawked at the two fearless femmes as they finished their early morning bath without being devoured by the ravenous reptiles.

The following day we had reached Epena with no further delays. Our Land Rover was waiting for us with enough fuel to take us back to Impfondo. With Liz and I sitting in the back of the Land Rover with all our equipment and supplies, we were making good time on the dark road back to Impfondo when a loud bang shook our vehicle and brought it to an abrupt stop. The left front tire had been shredded by some wood, freshly cut from the forest and left carelessly on the side of the road. Tom approached the nearest hut, shouting in Lingala and demanding that the owner of the wood attend the scene of the accident. The man stepped out of his hut, and did not seem at all concerned with Tom's protestations that the vehicle could have overturned with resulting fatalities. Even our promise to take the matter to the police did not ruffle him. "So what?" he spat. "The police will do nothing. You *mondelis* (whites) have no rights here." With that, he returned to his hut, leaving us to change the tire and continue on our journey to Impfondo. A cold shower, good food, and fresh sheets on comfortable beds reminded us to appreciate the simple things in life. In two days our flight would arrive to take us back to Brazzaville. The weather had suddenly turned cold and a huge thunderstorm was sweeping across the Oubangui River towards us. The solid wall of tropical rain raced towards the mission station while brilliant flashes of lightning lit up the sky. Ear splitting peals of thunder literally shook the ground as people raced for cover in their huts. As I watched the storm from within the mission guest house, I noticed Liz standing on the porch, filming the storm! This woman knew no fear.

As the morning sun rose over the forest, I awoke feeling ghastly, as though a flu-like virus had penetrated my body's defenses. Liz suggested that I had the same minor virus that had dogged her for several days before the expedition. The plane had finally arrived, and promptly suffered from a technical problem related to its landing gear. The pilot and a local French priest toiled away for most of the day to try and fix the problem with basic hand tools, but the task defeated them and an SOS went out to Brazzaville for help. The following day a small twin-engine Cessna arrived with a French technician who toiled almost all day in the hot sun to repair the problem before flying back to Brazzaville.

During our return flight I was beginning to feel worse by the minute. As the plane slowly made it way back to Brazzaville, I was freezing one minute and burning up with fever the next. I began to suspect that the "virus" I had picked up was much worse than I thought. Three hours later we thankfully landed at Brazzaville's

Maya-Maya airport. Gene Thomas had arrived back into the Congo a few days before and was there to meet us at the airport. Liz and I had originally planned to have dinner with the Thomases before flying back to Britain the following morning, but the delay in Impfondo left us scrambling to catch our Aeroflot flight back to Moscow, which was an hour away from take-off. In spite of the joy that Gene and I shared at meeting again after six years, he looked at my condition with concern and said, "You have malaria." After 42 years in the Congo, Gene Thomas could recognize all the symptoms of the dreaded parasite. I simply was not fit to travel, and if I had attempted to leave the Congo that day, I would have been unconscious by the time we had arrived in Moscow, and probably much closer to death's door. After seeing Liz safely to the departure lounge and on her way back to England, Gene rushed me back to the small two-bedroom apartment that he and Sandy shared near the airport.

Sandy immediately dispatched Gene to the nearest pharmacy for anti-malarial medication and sent me straight to bed. By this time I was beginning to sweat profusely again and suffer from delirium. I felt as if I was about to die. I had suffered from a few ailments and conditions in Africa, such as heat exhaustion, staff, amebic dysentery, and a mild bout of cholera, but malaria was simply dreadful.

Sandy re-appeared 30 minutes later with the biggest hypodermic needle I had ever seen. She injected a combined anti-malarial cocktail of Quinamax and Heptamyl straight into my tender gluteous maximus. I received the same daily injection every day for almost a week until the fevers had broken and the worst was over. I came to call this particular delight "Sandy's Special." In spite of the pain, it worked wonderfully. The malaria had left me weak for the remaining seven days I spent with the Thomases, but I was able to venture out with Gene on trips to the bank and to the Aeroflot office to confirm my airline ticket. On Monday, November 21, I said my farewells to Sandy, and Gene drove me to the airport. After checking in at the Aeroflot desk, I was directed to a sweltering departure lounge with no air conditioning, to await boarding. After a brief walk in the blazing sun, I finally boarded the Tupolev 154 airliner that would (hopefully) carry me back to Moscow without a hitch. During the flight I struck up a conversation with the young Congolese man in the seat next to me. He was returning to Moscow University where he was training to be a physician. Unfortunately, the Congolese government had run short of funding and the training for all future Congolese student physicians was limited to one year. With a population of only 2.5 million people, and in a country suffering from a HIV epidemic, the lack of proper training and resources for physicians and nurses will prove to be a major catastrophe. We parted company at Moscow, he to complete his medical "training" and I to the final leg of my journey. The flight to London was onboard the Ilyushin Il-86, a wide bodied, four engine jumbo that first entered service in 1980 to concur with the opening of the Moscow Olympic Games. Built to rival the Boeing 747, the Ilyushin never achieved the same success as its gas-guzzling, noisy Kuznetsov NK-86 low-bypass turbofans severely limited its range, servicing only with Soviet, East European, and Chinese airlines.

The 3½-hour flight to London was beautifully smooth, but a cold and rainy November soon had me running for a cab the minute I had passed through customs. Cold, tired, and just wanting to flop into bed, I shelled out $60 for a taxi to whisk me back to my Victorian terraced house in Gillingham. My wife and son were still in Mauritius and would not be back in England for another week, leaving me to grab a few basic grocery items from the corner store and gather up the small mountain of mail that had accumulated inside my front door.

Later that evening, as the rain and wind pounded my front window, I sat by the gas fire in my living room, sipping a hot mug of tea, and pondered the malaria parasite that could easily have killed me. References to the repetitive fevers of malaria are found as far back as 2700 BC in China during the Xia Dynasty. The term originates from the medieval Italian *mala aria*, meaning "bad air," as it was thought that the disease was contracted by breathing in the pungent air of swamps. However, in the 1700s, the natives of Peru introduced Jesuit missionaries to Cinchona bark, from which the active principle quinine was derived as an effective medicine to counter the then-deadly parasite. In 1880, a French army doctor by the name of Charles Louis Alphonse Laveran observed the malaria parasite inside the blood cells of malaria sufferers in Algeria, and for the first time this protozoa were identified as the cause of the disease. For this and later discoveries, he was awarded the 1907 Nobel Prize for Physiology or Medicine. Although modern medical advances in the development of effective treatments have saved countless lives, between 350 to 500 million people are still infected by the malaria parasite every year, resulting in between one and three million deaths annually. This represents one death every thirty seconds around the world, with children under the age of five being the most vulnerable group. By comparison, 2.5 million AIDS-related deaths were recorded worldwide in 2006, but Africa's biggest killer remains malaria.

A few weeks after my return to the United Kingdom, a few magazines, including the wonderfully entertaining *Fortean Times*, published various accounts of my second Congo adventure. The usual radio interviews were given for various BBC regional stations, and so I had made my mark on the world of cryptozoology, especially as a researcher of African cryptids, *mokele-mbembe* in particular.

One year later, in 1993, Tommy Boyd, a presenter on the British children's television show, *Magpie*, organized his own expedition to Cameroon in search of *mokele-mbembe*. The expedition was to be led by adventurer Barry Marshall. Mark Rothermel and I met Barry at Mark's home in the historic town of Hornchurch, Essex, to discuss his plans. We found Barry to be highly motivated, and he revealed that his eventual destination was to be the Mamfe Pool, located on the Manyu River in Northern Cameroon, where explorers Ivan T. Sanderson and Gerald Durrell encountered a huge roaring monster partially submerged in a flooded cave in 1932. As far as I knew, no further expeditions had penetrated that area in search of *mokele-mbembe*, and the only other report of any similar animal from northern Cameroon was recorded in 1948 at Barombi Mbo, the crater lake near Kumba. Indeed, Barry's research was based solely on an event that had occurred 59 years earlier and in an area that is now more heavily populated by humans. The recent construction of suspension bridge almost directly over the caves at the Mamfe Pool had greatly increased the human foot traffic in the immediate area and would almost certainly have driven away any shy, semi-aquatic animal to a more tranquil location well away from bothersome humans and herbivorous hippos. Mark and I discreetly exchanged doubtful glances and wished our fellow explorer the best of luck. News eventually reached us that Barry and his ten-man team had made it safely to the region and conducted enquiries among the people there. Unfortunately, their enquiries regarding large water monsters drew a blank, and the name *mokele-mbembe* was completely unfamiliar to them. I imagine they would have had better luck if they had tried *m'koo-m'bembo* or *jago-nini*. The expedition quickly gave up on their search and spent the rest of their time whitewater rafting on the rapids.

The next two years passed uneventfully, but not without another expedition in the planning. Terri and I discussed our future and what opportunities might lie ahead for us both, but especially for our son, Matthew. We both felt it was time to settle in pastures new, and the idea of immigrating to a new country became a more attractive proposition by the day. Terri had a number of relatives in Canada who spoke of a vast, modern, clean country with a small population and golden opportunities for all. Terri had excelled as a nurse, and I had entertained the idea of seeking a long-term career as a radio broadcaster and writer. We decided to take the plunge and applied for emigration to Canada. Within two years, we had completed all the necessary steps to qualify as Landed Immigrants for the "Big Country," and on March 30, 1994, we landed in Toronto, Ontario, to begin our new life in North America. For the first few months we lived with Terri's cousin, her husband, and young daughter, until we found a suitable apartment of our own. Only one month after arriving in Canada, our second son, Andrew, was born on May 5th at Scarborough General Hospital. Our first real Canadian in the family had arrived!

Just as we were settling into our new lives in Canada, the world reeled in shock at the massacre of over 800,000 ethnic Tutsis and moderate Hutus in just three months during the Rwandan civil war. Paul Kagame, leader of the Rwandan Patriotic Front, had seized power and overthrown the Hutu government. The point-blank refusal of the United Nations to intervene in strength, coupled with the virtual silence of the former colonial powers of France and Belgium, encouraged Kagame's followers to indulge themselves in an unbelievably savage bloodbath, creating a massive a flood of refugees that poured into eastern Zaire (now the Democratic Republic of the Congo, or Congo-Kinshasa), causing further mayhem by sparking the first Congo wars from 1996 to 1997, when Africa's most corrupt dictator, Mobutu Sésé Seko, was overthrown by Laurent Désiré-Kabila and his rebel army, backed by Angola, Uganda, and Rwanda. Yet another civil war erupted in the Congo from 1998 to 2003. Kabila himself was assassinated in January 2001 by one of his bodyguards, no doubt arranged by those who grew tired of his duplicity and increasingly dictatorship-style of leadership, while stalling the long promised multi-party democratic elections. Trouble too, was brewing in the former French colony of the Congo Republic, (formerly the Peoples Republic of the Congo). Former military ruler Denis Sassou Nguesso was unhappy at being an ex-dictator, and used the impending presidential elections of October 1997 to overthrow the democratically elected president, Pascal Lissouba, and seize power with the help of Angolan troops. The four-month civil war resulted in more than 10,000 deaths in Brazzaville alone. Although the war in the Republic of Congo ended in October 1997, shooting and other acts of violence between elements of the Congolese military and paramilitary groups was particularly fierce in some areas, particularly in the Pool Region, southwest of Brazzaville. In August of 1998, militiamen loyal to the former government launched a guerilla war against President Nguesso.

Government force eventually routed the militia loyal to ousted President Pascal Lissouba and his Prime Minister, Bernard Kolélas. Much of Brazzaville had been heavily damaged during the murdering rampages of the two main militia groups laughingly called "Cobras" and "Ninjas." Eventually, both sides agreed to a ceasefire in December 1999, with Lissouba eventually seeking exile in London, where he is still living, at the time of this writing.

The Congo, it seemed, was destined for disaster. It mattered not whether it was the Belgian side of the river or the French side. Presidential elections were all too predictable, with voting always ensuring that the biggest tribe got their man into |the top job. Murder, violence, and general mayhem always preceded and followed elections results. After the disastrous rule of the Belgians came to an end in 1960, the new independent government of Prime Minister Patrice Émery Lumumba lasted only ten weeks before he was imprisoned and later executed during the Congo Crisis (1960-1965), after which Lt. General Joseph Mobutu seized power. He later abandoned his Christian first name, adopting the African name of Mobutu Sésé Seko Nkuku Ngbendu wa Za Banga ("The all-powerful warrior who, because of his endurance and inflexible will to win, goes from conquest to conquest, leaving fire in his wake and arising from the blood and ashes of his enemies like the Sun which conquers the night."). In spite of the ridiculous name, Mobutu was a typical African dictator, robbing his people blind and amassing a fortune safely stashed away in various Swiss bank accounts. He murdered most of his political rivals, and regularly railed against the USA, Belgium, and France (though still accepting foreign aid from them), and had his image printed on all Zairian currency. Upon his death in 1997 from prostate cancer, Mobutu's personal fortune was estimated at five billion US dollars. The man who once lived in sumptuous palaces dotted around the former Zaire, and who flew to Belgium in his private airline for dental appointments, died in exile in Morocco with few to mourn him. In spite of his self-appointed god-like status, Mobuto was not that different from other African dictators who have left behind legacies of torture, mass murder, and corruption. The real tragedy is the former Belgian colony still has vast reserves of gold, copper, cobalt, rubber, and diamonds that could turn the country into an economic powerhouse. The Congo River has the potential to provide enough electricity to light up half the continent, but there will be very little investment in the rebuilding of the Congo until the vicious circle of civil wars and mass-murdering dictatorships comes to an end. Yet, in spite of its problems, the Congolese people remain friendly and accommodating to those visiting their land.

With two expeditions under my belt, I gave much thought of returning to the Congo. I was convinced that *mokele-mbembe* was a living animal, but its apparent rarity and semi-aquatic habits made it a very elusive quarry indeed. Trying to find such a creature was not only proving to be very difficult, but the fear that the animal had instilled upon the inhabitants of the Likouala region, coupled with their cultural and superstitious beliefs and traditional mistrust of white outsiders made such a task truly daunting, to say the least. By 1995, I had acquired my first computer and printer, thus gaining access to the wonders of email and the ever-growing world wide web. In 1999, I had been contacted by Kent Hovind, a creationist speaker and debater from Pensacola, Florida. He was fascinated by the possibility of living dinosaurs and offered me the opportunity to co-author a book with him, which would be published through his organization, Creation Science Evangelism. While I did not agree with everything that Kent used in his creation presentations, I did not want to miss the opportunity of participating in such a project, and we started almost immediately. The finished tome was entitled *Claws, Jaws & Dinosaurs*. The book was devoted to cryptozoology for children and was the first book as far as we knew to be devoted to cryptozoology from a creationist perspective. The sales were facilitated through various creation science organizations and the giant online bookstore, Amazon.com, but the attention that the book received solicited a number of invitations from both secular and Christian radio and television stations requesting interviews. Invitations to speak on the topic of cryptozoology in general and living dinosaurs in particular came from all over North America. Interest in *mokele-mbembe* was again beginning to mount.

In November 2000, Adam Davies, an intrepid globe-trotting young English explorer, ventured alone to the Congo and teamed up with Marcellin Agnagna for a trip to Lake Tele. Agnagna hinted that Lake Tele was not the best place to see the mystery animal, and instead pointed to a smaller lake on his map called "Lake Makale." In spite of this, Agnagna, Davies, and two guides, Sam and Sylvester, still trekked to Lake Tele, but no *mokele-mbembe* was observed. Davies made enquiries about the mystery animal among the Aka pygmies that lived near the mission station in Impfondo. He received some intriguing information from them through the resident missionary Paul Ohlin, who acted as a translator. At the village of Boha, Davies again received further information from a village elder who had seen *mokele-mbembe* many times. He drew a picture of the animal in the soil on the ground, revealing a distinctly sauropodian outline. Again, the verbal description was of a bulbous body and long, giraffe-like neck ending in a small head. The informant also mentioned that the male possessed a spike or horn-like protuberance on its head which the female did not, which corresponded well with the two long-necked animals observed in Barombi Mbo in northern Cameroon in 1948. But the informant could have easily confused *mokele-mbembe* proper with the *emela-ntouka*. Time ran out for the expedition and Davies had to abandon his desire to visit Lake Makale. I hope one day he will return for another crack at finding *mokele-mbembe,* and this time at the mysterious Lake Makele—if it really exists.

Lake Tele

Likouala Swamps at Sunset

> The reason I wrote the book (*Les Derniers Dragons de Afrique*) was because people were looking for them in all the wrong places.
>
> —Dr. Bernard Heuvelmans, in a letter to the author on the search for living dinosaurs in Africa.

6

Target—Cameroon!

With almost twenty expeditions having searched unsuccessfully to find *mokele-mbembe* since 1980, I decided to return to Africa for a third crack at this perplexing enigma. The prospect of conducting a third expedition to the Congo had become less attractive, as the continued instability of the country made it dangerous to foreigners and even more expensive than before. After digging out my old maps of the general region of the Congo Basin, I began to look up the older historical records that referred to *mokele-mbembe* and other very similar animals. Apart from a few odd reports collected by James H. Powell, Jr., in the late 1970s concerning the *n'yamala*, which was reputed to inhabit the Ogooué and Ikoy Rivers in Gabon, almost all other reports stretched back to the 1950s or earlier from various other locations as far away as Zambia. However, the 1913 report of *mokele-mbembe* written by Freiherr Von Stein Zu Lausnitz placed the geographical location of the animals in the rivers of Cameroon, quite close to the modern-day Congo Republic. My maps further revealed that southern Cameroon bordered the northern Congo, which could be used by *mokele-mbembes* to move freely in a large area that included deep broad rivers flanked by virgin forest with a very sparse human population and very little river traffic. "Perhaps this would be an ideal location for further research," I thought. And with that very thought, I began to draw plans for a small investigative expedition into southern Cameroon.

Although I preferred to work through the offices of various missionary organizations in Africa, I did not know of any missionaries in Cameroon who could help, but a quick search on the Internet quickly located a missionary organization called World Team that had at least two families of missionaries working in the country. Better still, the headquarters of World Team was located in the small city of Mississauga, which is now part of greater Toronto. I telephoned the organization and asked to speak to whoever oversaw their fieldwork in Cameroon, and to my surprise, their Field Director for Africa was a fellow Scot named John Wilson. After explaining the nature of my field research, John kindly forwarded the email address of Claude and Jennifer Daouste and Phil and Reda Anderton, two missionary couples who were engaged in evangelical and medical work among the Baka pygmies. I sent an email to Phil Anderton explaining my previous fieldwork in the Congo regarding *mokele-mbembe* and my desire to continue in Cameroon. Did he know of anyone locally who could assist me in traveling into the interior and interviewing the natives? A few days later, Phil responded and mentioned that Pierre Sima, a local plantation owner and elephant tracker was intrigued by my search for *mokele-mbembe* and would travel into the southern area of the country to make enquiries among the people there. Two weeks later I received a second email from Phil, explaining that Pierre had traveled to the border with the Congo Republic and spoken with many of the villagers who lived near the Boumba and Dja rivers. To my delight, Pierre discovered the people were very familiar indeed with *mokele-mbembe*, but known to them in the Baka (pygmy) dialect as *la'kila-bembe*. The descriptions given to Pierre by over a dozen eyewitnesses matched those by the Lingala-speaking tribespeople of the Congo: a long thin neck, a body sometimes bigger than an elephant, the small snake-like head, a long flexible tail, and a row of rigid dermal spikes that adorned the male of the species. The mention of the dermal spikes caught me by surprise. This was a physical feature of some sauropod dinosaurs that was unknown to modern-day paleontologists until 1991, when a fossilized diplodocus was discovered in Texas with beautifully preserved skin impressions upon the rock, including clearly defined dermal spikes. Yet the same feature had been attributed to the *la'kila-bembe* by various eyewitnesses whom Pierre had interviewed.

Pierre's report certainly convinced me that a short trip to the river system of southern Cameroon was essential if *mokele-mbembe* research was to continue without being hamstrung by corruption, mountains of red tape, and civil war. Barely weeks after receiving Pierre's report, an email arrived from David Woetzel, a businessman in Concorde, New Hampshire. David had a science degree in physics from Bob Jones University, was a creationist, and would be interested in participating in an expedition if one was being arranged in the near future? His timing couldn't be more perfect, and I quickly replied in the affirmative concerning the proposed Cameroon trip. After exchanging a few brief emails, I discussed my plans at length on the phone with Dave, who suggested we meet in Canada to finalize our plans for the trip.

Two weeks later, Dave drove to Toronto with his lovely wife, Gloria, a native of Columbia, and their children, Jonathan and Heidi. My own family accompanied me to Toronto where we met with the Woetzels in the front lobby of their modest hotel. We quickly agreed to convene for lunch at a nearby McDonald's restaurant, where

Terri and Gloria hit it off almost immediately, and the kids focused on the two PlayStation games in the restaurant, while I gave Dave an overview of my research to date, including a selection of photographs from my two previous expeditions. The meeting did the trick. Relieved to find that I was not some deranged lunatic chasing after mere fables, Gloria felt more reassured that her husband was teaming up with a regular, down-to-earth family man with a streak of adventure like himself. The last few weeks were spent collecting various camping items, tents, sleeping bags, a cooking set, first aid kit, and rations. Dave very generously provided the airline tickets and purchased an inflatable boat for river travel and a second tent with a sleeping bag for Pierre.

Finally I prepared a flip chart with images of various kinds of animals. These included the brown bear, the elk and the beaver. The second set of images included known African animals such as the elephant, hippo, gorilla, chimpanzee and hyena. The final set of images concerned dinosaurs and other fossil reptiles, including a *Tyrannosaurus*, *Diplodocus*, *Iguanodon*, *Stegosaurus*, *Apatosaurus*, and *Triceratops*. I also included a drawing of the *kalinoro*, a small bipedal primate with long, quill-like hair that allegedly inhabited the forests of Madagascar. There had been a few vague reports of odd-looking primates living in the Congo, so it would be interesting to see what the Baka would make of these images.

During our preparations for Cameroon, I was contacted almost out of the blue by Greg Richardson, a Christian business entrepreneur in New Jersey who headed a company called Viable Ventures. Greg was keen to support our efforts in locating and filming a specimen of *mokele-mbembe*. Although Dave owned a video camera, Greg felt that it was important for us to have two cameras and kindly sent me his own Canon camcorder with carry case, extra film, and lots of spare batteries, along with $800 to help with expenses in Africa. All that was left was to secure our visas (mine from the Cameroon High Commission in Ottawa) and a supply of Lariam malaria pills, which are essential for any visit to Equatorial Africa. John Wilson kindly provided me with some maps of the area we intended to explore, a mosquito net for protection against the potentially deadly insect, and a laptop for Claude Daouste, sent back to Canada for repairs and ready to be returned to him in Cameroon, as there were no computer repair facilities anywhere in that country.

On the last few days prior to our departure, I was hit with a sudden bout of influenza, which was quickly followed by an attack of bronchitis. My misery was further compounded by a painful ulcer on the roof of my mouth which made eating and drinking almost unbearable. The next two days were spent in bed to recover with rest and medication. My departure for Africa was a mere 48 hours away, and my condition worried Dave, who thought he might have to travel to Africa alone. In the end, Providence intervened and my condition improved dramatically. On November 3, 2000, I boarded an American Airlines flight to St. Louis, where Dave was waiting for me in the Air France departure lounge. Together we poured over maps, discussed Pierre's findings, and focused on the trip ahead. Our conversation of pygmy tribes, unexplored swamps, and monsters lurking in the dense recesses of Africa seemed to amuse of some of our eavesdropping fellow travelers. We were more than ready for the final leg of our journey. As the Air France Airbus A320 whisked us to Paris at 37,000 feet, we tucked into fine French cuisine and a glass of wine. The painful ulcer in my mouth diminished rapidly by the time we entered North African airspace, which allowed me the simple pleasure of enjoying a decent cup of genuine French coffee.

On November 4th at 6:00 p.m. we landed at Douala, the port city of Cameroon, to be greeted by the cloying heat of the African night. Shortly after the passengers bound for Douala had left the plane, the Captain announced that the plane had developed a technical problem and remaining passengers who were traveling on to Yaounde would also have to disembark until further notice. Air France arranged for a reasonable dinner and free drinks to be served in the airport restaurant while we waited for our plane to be made ready for the last leg of our journey. Unfortunately the delay lasted for six hours in a single terminal airport with non-existent facilities, evil smelling toilets, and rows of public telephones that did not work. Finally, we boarded just after midnight and arrived in Yaounde at 1:00 a.m. to be met by Walter Loescher, an American missionary based at the Cameroon Bible Institute, and Pierre Sima, who was visibly relieved that we had made it to Yaounde safely. Our problems were still not over as we discovered that Walter's Toyota truck had developed a flat tire in the airport parking lot. After much physical effort, we managed to replace the deflated tire with a spare and made it to our hotel by 2:00 a.m. The rooms were shabby; the bare porcelain toilet bowl had no seat, the toilet tissue holder was empty, and the shower only ran cold water. Dave and I were too tired to care and literally flopped into our beds, exhausted but happy to be at the end of our long journey.

The following morning, Pierre arrived at our hotel in a Toyota mini bus, which seemed to be the preferred vehicle for the majority of Cameroonians. After lunch with the Loeschers at the Cameroonian Bible Institute, we exchanged US dollars for the familiar, colorful CFA currency with Walter, then embarked on another six-hour journey to Bertoua, a large town west of Yaounde where Pierre maintained a plantation. Once we were out of the city, the roads became little more than muddy tracks full of rickety, makeshift bridges and large sinkholes filled with filthy brown water, so typical of Equatorial Africa. The Toyota became bogged down on several occasions, which forced us to push, pull, tug, and heave our stricken vehicle out of the muddy pools. In spite of our lack of sleep, we somehow managed to keep up this exhausting routine for 340 kilometers until we finally reached Bertoua. As the sun dipped below the distant forest, we arrived at Pierre's plantation in Dimako, just outside the town of Bertoua. Filthy, exhausted, and covered in the iron-rich red earth that is so common in Cameroon, we introduced ourselves to Claude and Jennifer Daouste, the Canadian missionaries who were working with the Baka, translating their language and developing a culturally appropriate church. They certainly had their hands full, with four children, April, Bethany, Julia and Geoffrey, and a full-time ministry to boot! It was a relief to deliver their laptop—and ourselves—in one piece. After a shower, change of clothes, dinner, and a decent night's sleep in a small, single building, almost like a bachelor unit for single missionaries, Dave and I felt like new men.

Claude rose early the following morning and treated us to a welcome cooked breakfast, which helped us enormously to prepare us for another very busy day ahead. Pierre arrived with a driver, Victor, in a 4x4 double cab Toyota truck, and whisked us back to Bertoua, to purchase our supplies for the trip to the south of the country. Bertoua is the capital of Eastern Cameroon with a population of 173,000 and growing. Later in the day we learned that George W. Bush, the former Governor of Texas, was now the President of the most powerful nation on earth and leader of the free world. Pierre gave us a quick tour of his home in Bertoua, a neat and tidy bungalow-style dwelling, before we stopped for lunch with Phil and Reda Anderton, American missionaries who were working in partnership with the Daouste family. Each family would spend a few weeks in the bush, then another couple of weeks in the town, alternating to keep the ministry fresh and growing. Reda was a physician who ran a free clinic for the local people at Pierre's plantation in Bertoua. She was also a full time mother to five children: Nathaniel, Nicholas, Naomi, Noah, and Nelson.

Pierre had visited with the Governor of Bertoua, Tantitiku Bayee-Arikai Martin, who kindly provided a letter granting us permission to travel freely throughout the Bertoua region, complete with his signature and official stamp. This official communication allowed us to pass through all police and military checkpoints on our journey without delay and free from having to pay bribes. One police officer, who was clearly dismayed at reviewing the governor's letter, admitted that he would have forced us to hand over a considerable amount of money. Corruption was now a way of life in almost every part of Africa.

At 2:00 p.m., we eventually climbed into the Toyota and headed south. As the journey progressed, the scenery became transformed into a surreal, primitive beauty. The hard packed dirt track continued to take us deeper into the sparsely populated south. The occasional small village broke the monotony of the 200 km journey to the town of Yokadouma, which means "where the elephant does not fall." Our accommodation for the night was a room with dirty tiled floors, a double bed with worn sheets, and pillows underneath a single bare light bulb. A plastic pail full of cold water was provided for bathing, but we were far too tired to freshen up and went to bed almost immediately.

Up by 5:30 a.m., Pierre took us to a small, simple café for breakfast where we dined royally on omelets and hot coffee. The morning was cold, and the town was shrouded in a mist that drifted in from the forest. Stores were already opening and large trucks hauling timber out of the forest were already rolling into the town. By far the most prosperous people here were the Middle Eastern traders with large open stores selling television sets, radio sets, and videocassette recorders. Yokadouma was becoming a popular stop-off point for organized tours visiting the rainforest and a major trading center due to its close proximity with the Central African Republic. But we were on our own mission and were anxious to continue with our journey. After Victor had filled up our Toyota at the local gas station, Pierre took a quick photograph of Dave and me next to a statue of an elephant in the town center before we continued on our journey south.

By noon we had reached the Baka settlement at Welele. Pierre introduced us to the chief, Timbo Robert, and the dozen or so immediate and extended family members who lived in the collection of mud huts. The group used to live in the forest, but the government relocated them to camps at the site of the road in order to facilitate the logging industry. After a lunch of bread, canned tuna, and oranges, we selected a suitable clearing on the opposite side of the road and pitched our tent before returning to the camp. It was time to find out just how much the Baka knew about the mysterious *la'kila-bembe*.

I opened the binder containing the flip chart I had previously prepared in Canada and allowed our hosts to look at the various images before them. The North American animals such as the elk, brown bear, and beaver drew a blank. After quickly conferring with the Baka, Timbo stated simply that none of those animals were familiar to him or any member of his village. I then showed them images of well-known African animals, such as the elephant, hippo, gorilla, chimpanzee, and leopard. These were all readily picked out as animals that were well known to the Baka. This simple test assisted us greatly in establishing a measure of accuracy in the Baka people's knowledge of the wildlife that inhabited the surrounding forest and river system. Then came images of presumably extinct dinosaurs. The *Tyrannosaurus rex*, *Stegosaurus*, *Pterodactyl*, *Iguanodon*, and *Ankylosaurus* drew the same lack of response as the North American animals. The Baka were simply unfamiliar with them and had never seen anything like them before. However, as soon as we showed them pictures of the sauropod dinosaurs, such as the *Diplodocus* and *Apatosaurus*, our hosts became excited and pointed at the images before them, speaking rapidly to Pierre and to one another. "*La'kila-bembe*," stated Timbo matter-of-factly.

"The Baka say that this animal lives here… in the river," stated Pierre. Dave and I were elated; we knew we were on the right track. As we continued to question our hosts in order to extract more zoological data on the animals, the details came thick and fast. The animals lived mainly in the river and adjacent swamps where they were observed browsing on the fruits and leaves. Timbo first observed one of the animals in 1986 when it abruptly surfaced in the middle of the Boumba River. It swiveled its head briefly as if to observe its immediate surroundings before plunging back into the murky brown depths of the river. Oddly, the illustration of an African native standing next to a hippo-sized *mokele-mbembe* on page 225 of Roy P. Mackal's book (*A Living Dinosaur? In Search of Mokele-mbembe*) was rejected by our informants. When I pressed Timbo to explain this, he stated that the animal in the illustration looked like *mokele-mbembe*, but was "too small." The animals that he and the four other eyewitnesses in his village had observed over the years were much larger, bigger than an elephant in most cases. Dave and I pondered on this question of the clear differences in size between the Congolese *mokele-mbembe* and the Cameroonian *la'kila-bembe*. The animals in Cameroon were similar morphologically, but much bigger. The body was more bulbous, the neck long and thin, ending in a distinctive snake-like head. The tail was long and flexible, and the legs were strong and positioned directly beneath the body, much like an elephant's. The animals were most often observed in the morning and later afternoon, usually browsing on the leaves and fruits of the overhanging tree branches at the river's edge. A few encounters had happened at night, but most Africans avoided the river after sunset. The Boumba River was a six-hour trek through the forest, and we were determined to get there as soon as possible. Timbo and his hunters would arrange to guide us the following morning and remain with us for most of our journey down the river.

The following morning we packed up our tent, ate a quick breakfast of coffee and granola bars, then prepared to head into the forest behind Timbo and his group of four guides. The trek was slow and tedious, as our guides had to hack their way through the thick foliage blocking our way as we attempted to follow the thin jungle path in the forest gloom. After three hours, we reached a clearing in the forest where we stopped for a welcome drink from our canteens. We were close to the river—the ground around us was covered with wood chippings from a tree that had been felled, and then hollowed out for use as a native canoe. Another hour of steady marching finally brought us to the river. The Boumba was perhaps up to two hundred meters wide and up to forty feet deep in places. The thick virgin forest hung over the river, which flowed steadily south where it merges with the Ngoko and the Sangha.

Timbo pointed to the spot in the middle of the river where he had observed the surfacing *la'kila-bembe* in 1986. He was astonished to see an animal so unusual abruptly appear from under the water. He stood rooted to the spot, as the animal seemed to briefly observe its surroundings before submerging back into the depths of the river. As we stared out at the river from the edge of the forest, a young boy named Shanga appeared at the far bank, paddling his canoe in our direction. Shanga was hired by Pierre to carry our supplies downriver during our expedition. Pierre explained that he had arranged for the canoe to facilitate our exploration of the river, but Dave and I preferred the comfort of our inflatable boat. Pierre decided to stick with the canoe and keep all our supplies with him while Dave and I floated down river in the inflatable boat. The Boumba was surprisingly quiet and tranquil; perhaps the best place to find any giant, water-loving herbivore. It was also the rainy season, and the deeper water and overhanging forest provided ample food and hiding places for our quarry. We drifted for hours in the hot sun, spotting only the occasional crocodile, or troops of monkeys capering along the high treetops just out of sight, wary of the humans below who hunt them for food with native crossbows firing poison darts. In spite of the hours we had spent on the river, we did not encounter a single canoe or observe any human settlements such as villages or fishing camps. Eventually, as the day began to cool in the early evening, Pierre called to us and gestured towards the north bank. Reluctantly, we steered our boat towards the bank and followed Pierre to a clearing in the forest where we made camp for the night. The cool damp night soon set in, but Pierre's culinary skills made up for the weather as we tucked into steaming hot bowls of rice, vegetables, and canned herring.

The two pygmies who joined us on the river trip started their own campfire, then spread out a reed mat on the ground and settled down for the night with their mighty hunting spears close at hand. The remaining three pygmies, including Timbo, decided to trek to our camp through the forest and do some hunting along the way.

By daybreak, Dave was already up and pumping fresh water into our canteens via his Katadyn filter, and what a marvel this little piece of equipment was! Like me, Dave easily fell prey to waterborne parasites, and decided to bring the Swiss manufactured mini Katadyn hand pump to purify our drinking water. The pump employed a long rubber tube with a cork float on the end, which would be placed into an available water source. The filter within could purify up to 2,000 gallons of water, removing all bacteria, fungi, and protozoa down to 0.2 microns, providing

Dave and Pierre at Camp

Baka Chief Timbo Robert, La'kila-bembe Eyewitness

pure, safe water. There is no doubt that this little eight-ounce technological marvel kept us free from falling sick, and continued to provide us with clean water for cooking and drinking during our expedition, a far cry from the waterborne parasites that crippled us on previous expeditions. After filling our canteens and water containers, we broke camp and prepared to head downriver. However, Pierre was worried that the ground team had not rendezvoused with us at the campsite, which delayed our departure for over two hours. Eventually we gave up and continued on our river journey. Dave remained as enthusiastic as ever and was keen to capture as much of our expedition as possible on his video camera. I shot a little film with the camera that Greg Richardson had lent me, but I reserved my film and batteries for that special moment when our quarry, the feared *la'kila-bembe*, would make an appearance.

After another four hours on the river, Pierre again signaled his intention to land on the east bank. Filbert, a plantation owner, was expecting us and welcomed us to his simple home, which comprised of a single hut and a shelter on sticks. Large baskets filled with coca beans attracted a swarm of honey bees, causing me to beat a hasty retreat to a less threatening location as I tried to catch up with my notes. Dave and Pierre headed into the forest to wash down in a pool of water, but ran into several columns of large ants. I later opted to freshen up with a large pan of water without the ants. The bees had been quite enough company for one day! Later, we changed into fresh clothes and prepared for dinner as two pygmies headed down to the river with our laundry. Once again we dined on rice, vegetables, canned herrings, and antelope, washed down by hot coffee. After setting up our tent on a slight slope at the edge of the forest, Dave began to ask our host about the *la'kila-bembe*. Had he ever heard of the animal? Had he seen one? No, he had not, but he did see the fearsome *dodu* in the heat of combat with a male gorilla. The sheer ferocity of this strange, bipedal ape as it fought to the death with a bull gorilla caused him to flee in terror.

The fearsome *dodu* gave even Pierre pause for thought. At 6'2" tall and powerfully built, he has explored more rainforest than any other Cameroonian, tracked elephants for weeks through dense bush, observed large groups of gorillas, stared down charging chimpanzees, and killed countless venomous snakes that attempted to sink their fangs into him. He had never before heard of the *dodu*. Even Pierre admitted that the Baka will go to places in the forest so deep and remote that even most fearless of Bantu hunters would never dare follow them!

Filbert mentioned that on the past few nights, he heard the repeated roar or bellow of a large, strange animal as it wallowed around in the river. He was sure that it wasn't a hippo, as they were never found in that stretch of the Boumba. He wanted to investigate, but was afraid in case the animal—whatever it was—was dangerous and attacked him or followed him back to his plantation, located very close to the river. There were many strange animals that lived in the river and the surrounding forest that the white man had not seen, and he did not want Dave or me to risk injury or even death by conducting our own nocturnal investigation. He repeated the same sentiments of old Chief Makoko many decades ago: "There is always trouble if a white man is killed." After we had pitched our tent and settled down for the evening, Dave continued to press our host about the *la'kila-bembe*. His eagerness for more knowledge about our ultimate goal made our Baka guides uneasy, and they asked Pierre to "advise" us to stop talking about it. Timbo later explained that the Baka believe that the *la'kila-bembe* has a "bad spirit," due to its ferocious nature. While they did not ascribe any supernatural powers to the animals, they simply did not like them, and avoided any contact with them lest they too lose their lives like the hippo and the elephant. It was a monster to be feared for sure!

Ever keen to record everything on film everything around him, Dave even filmed dinner, which was a sort of Cameroon-style "buffet" of rice, canned herring, and sitatunga (or marshbuck). Reluctantly, we retired to our tent for the night in preparation for the remainder of our journey in the morning. During the night, a tremendous tropical storm pounded the camp, sending torrents of water pouring down the slight grade where our tent had been pitched. Thankfully, Dave and I remained warm and dry throughout the night. The final leg of our journey continued with a slow drift down the Boumba, which was interspersed at places with small, rocky waterfalls. Our inflatable boat and wooden canoe was no match for the falls, and we circumnavigated these on foot as we slogged along narrow hunting paths near the river. Our hopes of exploring the smaller, promising Lapondji River were dashed when we simply ran out of time. This disappointed us greatly, as a pygmy fisherman by the name of Fazia who lived near this smaller river had informed Pierre of his own surprise encounter with a *la'kila-bembe* in the river. The man had paddled out into the river to check his fishing nets, only to find them snagged by some heavy underwater obstruction. As the fisherman heaved and tugged on his net in an attempt to free it from the mystery obstruction, he was suddenly thrown back as his canoe bucked violently from a sudden upsurge of water directly in front of him. Although Fazia barely managed to keep his small canoe from being capsized, he was astonished to see an enormous creature rise from the depths of the small river. He knew exactly what it was: the feared *la'kila-bembe*, the god of the river that no man could kill, and the killer of hippos and elephants. Even the biggest crocodiles would flee from it!

The description that Fazia gave to Pierre was by now thoroughly familiar to us: a long neck, a body as big as an elephant, the tiny snake-like head, and the long tail. Fazia paddled back to his hut as fast as he could without looking back, but he could hear the animal moving around in the river. He later thought that the animal had moved into the smaller river in search if its food supply, and to avoid being disturbed by the slight increase in river traffic as the plantation owners headed south to Moloundou with their vessels laden with bananas, mangos, cocoa beans, coffee beans, and manioc.

Our disappointment at not meeting Fazia or being able to explore the Lapondji River was heavily felt by Pierre, but time was simply not on our side. We were on a short two-week recon into a new location and our job at this point was to simply gather as much information as possible for a later, more elaborately-planned expedition. Our final bush camp was established by the river in the cool of the evening, as Pierre prepared supper and coffee. As we broke camp the following morning, we set off on the final leg of our mini-adventure, a solid, all-day trek through dense forest, broken by patches of open savannah. We rested mid-morning in a small hunting camp, where Pierre found a large marijuana plant. The weed was smoked by pygmy and Bantu alike, and was often traded in the villages. After a refreshing drink of water from our canteens, we continued on our journey, leaving the camp and the weed behind. By mid-day, Pierre discovered that a bridge linking a broad jungle path with a shortcut to the village of Mandele, our final destination, was still underwater and any attempt to continue on foot in that direction was simply too hazardous. Our only route was twice the distance through open savannah and the possibility of running into dangerous game, particularly wild elephants or aggressive buffalo. We had little option but to press on, regardless of the risks. Our route took us over a small, rickety footbridge that spanned a narrow fast flowing rapid of the Lapondji River. The "bridge" was little more than a few thin tree trunks that creaked under our weight as we crossed the rapids one person at a time. The Baka simply skipped across in spite of being laden down with our supplies. Pierre decided to wade across in the fast flowing waist-high rapids. Dave made it across the bridge safely, in spite of the fact that the wooden supports looked as though they were about to give way at any time. Finally, it was my turn. As I slowly negotiated the thin, slippery tree trunks, I lost my balance midway as the bridge simply disintegrated beneath me, sending me headfirst into the rapids. Although I managed to keep my head just barely above the water, my right foot was trapped under one of the logs. After much effort, I managed to free my foot and crawl out of the water with Pierre and Dave dragging me onto dry land. Thankfully Dave didn't manage to film my unscheduled bath!

Barely an hour later, we broke from the forest into open savannah. The ground was baked hard as the noonday sun beat down upon us mercilessly. There was no shade for miles as we slogged on along a narrow path that cut like a knife through the shoulder-high elephant grass. On several occasions we were forced to jump over thick columns of huge black ants—something that Pierre had forewarned us about. Dave and I tucked the bottoms of our pants into our thick, absorbent socks in order to deter the ants from sinking their formidable mandibles into our inviting white flesh. The Baka were not so lucky, having to beat off the tiny invaders from their exposed legs and feet. The heat was almost too much to bear, as my video camera, utility belt, and water bottle seemed to weigh me down on this brutal, final push to reach Mandele. The narrow trail through the thick foliage and tall elephant grass tore at our clothes and even pulled our shoes laces apart as we pressed forward in the blistering heat.

Finally we stopped to rest for a short time at a crumbling wooden bridge that had seen better days—I instantly passed out the second I hit the ground under the cruel, relentless sun. Dave, who never seemed to lack energy, found a nearby water source and filled our bottles and canteens with clear refreshing water mixed with orange crystals. After shaking me out of my unconsciousness, he presented me with a bottle of his special orange drink, which refreshed me remarkably quickly.

At 1:30 p.m. we pressed on, and finally reached Mandele at 5:00 p.m. The village was a stop-off point for the logging trucks, including white European visitors employed in the logging trade. Although we were filthy and exhausted, our appearance did not arouse much interest, for which we were grateful.

All we wanted was to rest in the shade for an hour, purchase some cold drinks, and be on our way. Finally, Victor showed up with the Toyota. Pierre paid off our Baka guides and we continued to Moloundou, stopping at a small river to wash, shave off our beards, change into clean clothes, and scrape the mud off our footwear. The pair of $8.00 sneakers I had purchased at a Toronto convenience store had held up remarkably well during our jungle trek!

The town of Moloundou and the *mokele-mbembe* encounter site at the ferry crossing were our final destination. The ferry across the Boumba River was little more than a large floating platform tethered to a couple of heavy-duty steel cables that spanned the river. There was enough room for one safari bus and about thirty passengers, who would be conveyed across the river by two men operating a large hand winch, which slowly moved the platform across the river. It was a laborious task, but considerably safer than attempting to cross in the narrow native canoes.

The ferry was not in operation during our visit, but Pierre pointed to a spot, roughly north of mid-river where two security guards observed a very large *la'kila-bembe* slowly making its way south on the river towards the ferry, only eight months prior to our arrival. The animal was unable to pass the ferry cables and turned around to retreat the way it had come. Badly frightened by the sight of the monster, the two young guards fled the scene, never to return. During Pierre's interview with one of the witnesses, the man described the animal as possessing a body as big as an elephant with a long, thin neck and a small head. The animal did not appear to be concerned with the cries of alarm that he and his colleague gave as the monster appeared in sight, but slowly turned around and headed north (upstream) after it encountered the thick steel cables that spanned the river. The river was little more than six feet deep at the time of the observation, thus forcing the animal to expose most of its body, neck, upper limbs, and part of its long powerful tail. The witness told Pierre that he and his fellow guard had never seen anything like the animal before or since. One of the security guards fled Cameroon and went to live in neighboring Congo, perhaps unaware that the river monster he had fled from was also well known there!

Later in the town, Pierre had left us with the vehicle as he went on an errand. Dave, as usual, whipped out his camera and started filming. Before I could caution him about filming in a town or city without official permission, a local *gendarme* showed up and immediately attempted to relieve Dave of his camcorder. A tug of war began between Dave, who was determined to hold on to his camera, and the gendarme who was determined to confiscate it. Just as I was beginning to have nightmare visions of spending years in an African jail for "spying," Pierre showed up and engaged the indignant police officer in conversation, pleading that we were mere tourists and not a threat of any kind to the town's security. As far as the police officer was concerned, Dave and I were "Europeans" who "owned" the forest, and loyal Cameroonians like Pierre had no business even befriending us, let alone upholding our innocence. Dave and I sat in the Toyota while Pierre continued to haggle with the indignant officer, who demanded Dave's camera and a 200,000cfa fine. Pierre eventually showed him the document from the Governor of Bertoua, allowing us to travel freely in his district.

After about thirty minutes, the police officer gave up and allowed us to continue on our way but not until we paid a 30,000cfa "fine." We reluctantly parted with the money and headed back to the road rather than lose any more time. The Cameroonians, like the Congolese, love their government documents, with colorful stamps, signatures and titles.

The road north was fraught with the usual deep muddy pools, rickety bridges and overturned logging trucks littering the road. At dusk, we stopped at a small village by the side of the road and pitched our tents for the night. The village children gathered around us and even followed us into the bush as we attempted to empty our bladders before bedding down for the night. As Dave remarked, the children didn't see many white visitors attend to their *toilette* and probably wanted to see just how white we really were!

Grateful to be in our tent for a good night's sleep, our hopes were dashed when we were awoken at 1:00 a.m. by a Baka woman arguing loudly with her husband. She had drunk a little too much palm wine and reacted violently when her husband rebuked her. Their argument then woke up a baby, which in turn began to cry, thus arousing the rest of the village. The drunken woman was then thrown out of her hut by her husband, but continued to berate him loudly, provoking half the village to get in on the act by beating the woman in an effort to subdue her into silence. After an hour or so, the melee subsided and we tried to get back to sleep, only to be awakened by the bleating of the village goats, which started at 4:00 a.m. and continued until we packed up and continued on our way. At one roadblock another police officer casually remarked that he would have demanded a considerable amount of money from us if we had not presented him with the governor's note. It certainly helped to have more official documentation that just a tourist visa when visiting Cameroon!

Although we remained exhausted from our jungle trek and lack of sleep, our day brightened slightly when we stopped at Yokadouma for a welcome lunch and to purchase gasoline for our truck. By 5:30 p.m. on November 14, we finally reached Dimako and Pierre's plantation where the mission houses were located. After a welcome shower and a simple but delicious dinner of pasta and tomato sauce prepared by Reda Anderton, we literally collapsed into a double bed with fresh sheets. We fell asleep around midnight, only to be rudely awakened by a thunderstorm that erupted at 6:00 a.m. and kept us awake for the rest of the morning. Would we ever get a decent night's sleep?

The following morning, Pierre mentioned that an elderly pygmy couple that lived on the plantation were very familiar with some of the mystery animals that lived in the Dja River region. After introducing us to Ebondo and Cephu and explaining the purpose of our visit to Cameroon, Pierre asked the elderly Baka couple if they would be willing to look at our binder of animal illustrations to see if they could recognize some of them. As with all previous informants, the couple quickly dismissed all North American animals as being unfamiliar to them. The African animals, such as the elephant, hippo, crocodile, gorilla, and chimpanzee were all quickly acknowledged as animals that they were familiar with. Finally, I presented them with the dinosaur illustrations. Almost all of these were again quickly rejected, with the exception of two animals. *"La'kila-bembe"* stated Ebondo, as he pointed to the picture of a *Diplodocus*. The old pygmy's eyes brightened as he recalled how as a younger man, he was fishing in the upper reaches of the Dja River when the sound of a large animal moving towards the river out of the forest caught his attention. Thinking that an elephant was approaching, Ebondo was astonished to see an even larger animal with a huge body, a long thin neck, and a very small head like a snake, push its way through the dense foliage and plunge into the river. He got a good look at the animal too, including its powerful forelegs that supported its massive body. He later described his encounter to the village elders who confirmed what he saw and advised him to stay away from certain parts of the river where the monsters were known to frequent. The second animal they both recognized was the triceratops. The *n'goubou*, the fearsome multi-horned killer of elephants, was sometimes observed in the savannah area between the Boumba and the Dja, but more often in the open savannah area between Cameroon and the Central African Republic. Dephu also remembered a time when she too saw a *la'kila-bembe* in the river, and another time when an *n'goubou* had been killed by a hunter with a rifle. She recalled observing the strange beaked mouth, neck frill, and several horns on the animal, which was as big as an elephant, before the entire village chopped up the animal for its meat. With this information, I thanked our informants for their patience with our questioning and left them to consider the information they had given us. Here was further eyewitness corroboration of two distinct animals that looked unmistakably like dinosaurs. Could we have really found a rare primitive paradise, a lost world of some kind where that last few dinosaurs had somehow survived into the twenty-first century? If this was truly the case, how much longer could they survive?

The following evening, Claude and Jennifer Douste arrived with their four children to continue with their fieldwork, while the Andertons moved back into the town on the two-week rotation. At 6:00 a.m. on the morning of November 16, Dave, Pierre, and I packed up our gear, ready for the five-hour drive back to Yaounde. Claude and Jennifer also arose early to send us on our way with a delicious, cooked breakfast. By 7:00 a.m., Victor arrived with the Toyota, fully fueled and ready to go. We reluctantly made our farewells and climbed into the truck. We waved at our hosts until we were out of sight and on the road to the big city. We would miss them.

Arriving back at the Cameroon Bible Institute at noon, we were quickly billeted in a comfortable guest house and enjoyed a large lunch with the Loeschers. Our expedition had officially come to an end after a 13-day recon with positive results. After an unsuccessful attempt to telephone Air France from the institute, we decided to play it safe and headed into town to confirm our flight back to the USA. In Africa, it was always wise to make absolutely certain that your seats were reserved, as they would often be taken by other passengers by the time you were ready to check-in at the airport.

Later that evening after dinner, Dave plugged his video camera into a small television in the Loeschers' house and ran the three hours of video footage he had shot when in the bush, giving our hosts a first-hand glimpse into the untouched forest and remote rivers. Claude, a Cameroonian pastor who worked among the tribes in the south of the country, became greatly intrigued by our quest and promised to keep his ears and eyes open for any interesting or unusual reports concerning our suspected dinosaur.

The remainder of our time was spend recovering from our jungle mission, and shopping for souvenirs in Yaounde, where I purchased two Cameroon National soccer outfits for my sons and some malachite jewelry for my wife. After a wonderful lunch with our hosts and founder of the institute, Dr. and Mrs. Fred Hocking, we returned to the guest house where Dave set up his video camera to record a discussion on our findings in the bush. We were both delighted with our brief but rewarding recon into what proved to be fertile *la'kila-bembe* territory. Our informants all not only picked out the sauropods unerringly as being closely representative of the mystery animal, but the animals themselves had been seen often and recently. The Baka pygmies also differed considerably from their Bangombe and Kelle counterparts from the Congo, in that they did not attribute any magical powers or spiritual beliefs to the animals. They did not like the animals because they were dangerous when approached, destroyed fishing nets, occasionally capsized canoes as they surfaced in the river, and disrupted the fishing activities and river passage of the natives. Furthermore, absolutely no one had ever questioned them on *la'kila-bembe* before. They had never been shown pictures that resembled the animals before, never been questioned about them before, and had never been asked to draw images of them in the soil before (as we had done). As far as we knew, Dave and I were the first white men since the Von Stein expedition of 1913 to have explored southern Cameroon specifically in search of *la'kila-bembe*. It was a great start for us and whetted our appetites for further research.

On November 18, Walter and Pierre drove us to the airport and made sure we checked in at the Air France desk, ensuring that our seats were indeed reserved and waiting for us. Our flight to Paris went smoothly, as did our connecting flight to St. Louis. It was there that Dave and I parted company as I ran to catch my connecting Air Canada flight to Toronto. A few weeks later, Dave sent me a videocassette entitled "Behemoth or Bust," a video diary of our trip to Cameroon. The film is still currently for sale on Dave's website, Genesis Park, which also has some of our expedition photographs for general viewing as a slide show. This rekindled interest in *mokele-mbembe*, which had waned somewhat in the late 1990s. I was confidant that not only was *mokele-mbembe* a living animal, but it was more widespread that

William Rebsamen

we had previously thought. The interconnecting river systems that spanned Congo, Gabon, and Cameroon certainly required considerably more exploration.

In early December 2001, I received a phone call from Jo Sarsby, a film producer for the BBC in Bristol, England. Was I planning another expedition to Cameroon in the near future?

Discovery Channel Expedition, 2001

Jo was intrigued by the *mokele-mbembe* mystery and was interested in the possibility of filming an expedition in pursuit of the animal. She had previously contacted another group in England that was also planning an expedition to the Congo, but they were still in the preliminary stages and had not yet secured their government papers, visas, or adequate funding. We discussed her proposal over the next few days and decided that just such a project might just be feasible after all. I now needed to organize our expedition team, find funding, and purchase all the necessary equipment and supplies. Could we get everything organized and ready to go within three months?

Encouraged by our findings in Cameroon, Greg Richardson decided to look for one or more sources of funding for our expedition. One such possibility came in the person of Paul Rockel, a Waterloo, Ontario, resident who founded Regal Capital Planners. Paul donated $1.5 million to two local Lutheran schools because of their track record of maintaining strong Christian values. After reading about Paul's donation in a local Ontario newspaper, I passed the clipping along to Greg, who in turn contacted Paul to discuss the possibility of financial sponsorship for our expedition. Paul became greatly intrigued by the possibility of dinosaurs still living in the remote swamps and rivers of West and Central Africa, and requested further information. After a few phone calls describing our past expeditions and future plans, Paul invited me to his offices to discuss the matter further. The meeting went splendidly, and Paul sent me on my way with a firm handshake and a solid commitment to funding our expedition. Now it was time to put together a new team.

There were relatively few fellow cryptozoologists who would be available to go on such a major expedition at such short notice, especially those who were at least sympathetic to creationism. Dave Woetzel had too many business commitments to travel again so soon, so I turned my attention to three fellow Christians and cryptozoologists with whom I had been corresponding for several years. John Kirk, a fellow Briton, was born and raised in Hong Kong to army parents, now living in Vancouver, and president of the British Columbia Scientific Cryptozoology Club. Scott Norman, a website developer and photographer from Fullerton, California, was my second choice. Like John, Scott had not only studied cryptozoology but was an avid field researcher and was very well known in the field. The fourth member of our team was Robert (Rob) Mullin, the youngest member in his 20s, from Manhattan, Kansas, also known as the "Little Apple." Rob was an avid cryptozoologist, a gifted science writer and debater in the seemingly endless battle between scientists who supported both sides of the creation-evolution controversy. Having quickly received confirmation from all three regarding their availability, I began to prepare photocopies of visa application forms and a list of personal supplies for all three team members. A list of much needed camping and field supplies was quickly drawn up, and most of these were purchased from a variety of camping and outdoor stores in Pickering, where I lived. Scott would bring a fish finder/sonar unit for river work, and Rob was given the task of collecting all the food supplies and other perishables needed for our base camp in Cameroon, wherever that might be. Finally, a local travel agent managed to secure four return tickets from Canada and the US to Cameroon via Paris.

In December 2000, Jo flew to from England to Toronto in order to meet with me to discuss our field itinerary and equipment. A single career woman in her thirties, Jo was dedicated to her profession and had considerable experience in filming in the remote wildernesses of the world, including Cameroon five years before, which was indeed a bonus for us. We discussed the inevitable contract which would seal the deal, and this would be faxed to me from Bristol once everything had been prepared for the trip. Everyone was excited at the prospect of participating in a major expedition that would eventually be seen on television all over the world. Was it now finally time to make the breakthrough in *mokele-mbembe* research and find our elusive water dragon at last?

After conferring with Paul Rockel on all our preparations, he committed $35,000 to cover all expedition expenses. I was speechless. What could I say to such generosity? Jo Sarsby secured a contract and additional funding from the Discovery Channel, and would be working with a cameraman and sound technician from New Zealand. On top of everything, the BBC had committed $15,000 in additional funding, which covered all our purchases, travel costs, and on-ground expenses in Cameroon nicely. I regularly conferred with Greg Richardson on the details of the contract, including the rights to any film or photographic evidence that we might secure during the expedition. Finally, Jo completed the deal by working out the fine print with Greg to everyone's satisfaction, and the contract was faxed to me from Bristol. Just two weeks before our departure, Paul Rockel called me and decided to increase his contribution to $50,000. It was almost too good to be true. But it was! Everything had been prepared by January 31, 2001. Everyone had secured their visas, received the required inoculations for Cameroon (yellow fever and cholera) and the plane tickets made it on time.

Scott would fly to Los Angeles LAX and meet with Rob, who had to take two domestic flights to arrive at the same destination. They would travel together on Air France to Paris where we would all meet up. On February 15, John Kirk flew from Vancouver to Toronto, a six-hour flight, and then met me at Pearson International Airport just outside Toronto.

John Wilson of World Team and his wife Gloria arrived at Pearson to see me off and take back the maps and the mosquito net I had borrowed from him for the previous trip. We would purchase new ones in Yaounde. After taking the additional step of having my suitcase and two large cartons of supplies plastic wrapped, I paid an additional $140 for the excess baggage just a John Kirk arrived.

Tall and dark with a commanding presence and an air of aristocracy, John more than lived up to his reputation as a world-class cryptozoologist par excellence. We stood and chatted for a time before we were finally required to pass through security prior to boarding our Paris flight. John and Gloria left with a promise to pray for our safety when we were in Africa, for which we were truly grateful.

In spite of the turbulence for most of the flight, John and I chatted about many things cryptozoological, including the expedition, which we were eagerly looking forward to. Finally we were able to snatch some sleep after dinner, eventually landing in Paris at 9:30 a.m. As our connecting flight would not be available until the following day, we booked two rooms in advance with the *Hotel Opera Cadet*, a centrally located hotel with friendly staff and pleasantly appointed rooms. For $120 US per night, we more than determined to make ourselves at home, unwind for an hour, and watch European soccer on television. John was a huge soccer fan who once played semi-pro and even interviewed the legendary Pelé for a Hong Kong television station where he was a producer before moving to Canada.

Rob and Scott arrived at the hotel two hours later. Scott wore an Indiana Jones style hat, not unlike the one I had purchased for the expedition and Rob was dressed almost entirely in US Army surplus clothing, which no doubt caused a few odd glances on his long journey. We later did a little sightseeing in Paris, traveling on the Metropolitan, the city's famed subway system. Paris has the most closely spaced subway stations in the world, with 245 stations within the 41 square kilometres of the city. The first line opened in 1900, during the Exposition Universelle world fair. The system expanded quickly until the First World War and the core was complete by the 1920s. Extensions into suburbs were built in the 1930s to cope with increased passenger use, which today stands at 4.2 million per day. Our final visit was to the

Eiffel Tower, but the cold February weather sent us shivering back into the Metropolitan and back to the hotel. Later that evening, we dined at a wonderful old restaurant where John and his lovely wife, Paula, had enjoyed a honeymoon dinner a year before. This beautiful old bistro was located in a beautiful old cobbled square Montmartre around the corner from the church of Sacre Couer where artists sketched portraits for tourists and violinists played around the dinner tables. Later, over coffee we discussed how we might work with the BBC, who would be waiting for us in Yaounde. Everyone was excited at the prospect of reaching Africa.

We arose early on the morning of February 17, to head off to the airport. The hotel provided a sort of buffet breakfast of cereal, croissants, fruit juice, and coffee, which we quickly ate before loading up a minibus taxi with our bags and boxes filled with expedition equipment and supplies. Arriving at Charles de Gaulle International Airport, we found ourselves in a line up at the Air France check-in that dealt with all African departures. We waited patiently for 90 minutes to be booked in, after which Air France tried to charge us an additional $2,000 US for our "excessive baggage," which in total weighed 690 lbs. John immediately went on the offensive with Scott in support, insisting that we had paid for the excess baggage in Toronto, and that we had expected to connect with our flight to Yaounde within a few hours after landing in Paris, rather than stay overnight in the city. After much haggling, Air France relented and booked in our baggage without further cost to us. The delay had cost us valuable time, and according to the departure screens, our plane was preparing to take off without us, which necessitated a rather dramatic sprint through the airport to the departure gate. To our great relief, our flight had been delayed by an hour, giving us an opportunity to unwind and relax on the plane prior to take off.

Although the Airbus A340 was quiet and comfortable, our journey was disrupted by a steady amount of turbulence almost all the way to Yaounde. Jo Sarsby was on the same flight, giving me an opportunity to introduce her to John. Eight hours later, Pierre was there to meet us at Yaounde with a minibus waiting to whisk us into town. Our boxes and large bags attracted the attention of the customs officials at the airport, who wanted to search every item of luggage. Once again John quickly dealt with the situation by searching out the chief customs officer and loomed over him, speaking rapid, but very fluent, French and waving our baggage itinerary list in him face explaining that we were scientists on an important mission to Cameroon. The man decided that rather than haggle with the 6'5" tall, worldwide chief of the clan Kirk, he would rather just let us go on our way. The sights and sounds of Equatorial Africa were new experiences for Scott and Rob, who took in their surroundings as we headed for the capital city with its bright lights and maniac traffic. The reasonably priced *Hotel Meumi* was comfortable but somewhat spartan, typical of Africa. Our film crew had arrived 24 hours before and later met us for dinner. Nigel, our sound technician, and Warren, our camera operator, were both from New Zealand and had worked all over the world. Unfortunately, Air France had lost or misplaced their luggage, so the crew was without a change of clothing. Many of the missionaries working in Cameroon referred to Air France as "Air Chance" due to their unenviable reputation of losing luggage with alarming regularity. Francis Ngwa Diba, a smartly dressed and intelligent young businessman "fixer" for the BBC showed up at the hotel with our government papers. The papers included a certificate from the Minister of Communication, which allowed us to film freely in the country without being arrested as spies.

February 18 fell on a Sunday. With little else to do, we stayed at the hotel for breakfast and lunch and later went to the Cameroon Bible Institute for the evening service. A number of new arrivals from the USA were living at the institute, many of them younger people on short-term mission service. Dr. Hocking kindly invited us to stay at the institute once our expedition was over, in order to help us save money on hotels and meals in the city. We finished our day with dinner at LGM, a French burger and ice cream parlor, which we would frequent during our stay in Yaounde.

The following Monday we were at last able to get organized for our departure. After acquiring the additional funds sent to us via Western Union by Greg Richardson, we then embarked on a shopping spree, purchasing three car batteries, a power cord, and additional batteries for our fish finder/sonar unit. Rob, it seemed, attracted the attention of every young, single Cameroonian girl regardless of where we went, whether it was a café, a store, or back at the hotel. Although John, Scott, and I were amused by the attention that Rob was attracting, I felt a little sorry for him, as his sole intention in coming to Africa was to participate in a major expedition and nothing else!

Our final day in Yaounde was spent purchasing some food items for the expedition, including a field cooker with a supply of kerosene fuel. Jo and her crew were beginning to get rather frayed around the edges as their luggage had still not been recovered, so they went off to the marketplace to purchase some items of clothing for the trip. Pierre spent most of his day in town and was having difficulty in finding suitable transportation for everyone. There were plenty of older vehicles around that we could hire, but the BBC preferred newer vehicles and proper insurance before embarking on the journey to the south, which was understandable. After a meeting with Joe, Nigel, and Warren, we discovered that we did not bring a copy of the contract with the BBC, which we needed to review our obligations regarding transportation. John and I telephoned Greg in the USA to explain the situation, and he faxed a copy of the said document to our hotel. After a careful review of the contract, it was clear that any final decisions on the provision of transportation and the safety thereof, lay firmly in the hands of the expedition team and not with the BBC. The following morning, Pierre found a Toyota mini-bus and Jo arranged for the hotel to provide a driver, Christoff, and an air-conditioned Toyota SUV for the journey to Bertoua, our next destination. Scott had brought along a satellite phone from the USA to send daily reports to Greg by telephone or email. We opted for the email feature, as it was cheaper.

On February 22 at 4:00 p.m., we left Yaounde in three vehicles: John, Rob, Scott, and I in the Toyota Minibus, the BBC crew would follow later in the SUV, and Pierre would travel in a large blue safari bus leased from *Alliance Voyage*, to carry our supplies and equipment on the dangerous, unpaved road south. After two hours into the journey our vehicle suddenly swerved off the road and came to a grinding halt. A quick examination of the problem found that the front left brake of our vehicle had literally disintegrated, leaving us stranded. The BBC, unaware of our plight, zoomed past us in their SUV. Fortunately, Pierre spotted us on the side of the road and stopped to pick us up.

The minibus would have to stay where it was, as we were behind schedule and needed to reach Bertoua as soon as possible. Fortunately, the safari bus could carry all four of us, and our supplies. With some relief we had made it to Bertoua by midnight and booked in at the Manga Hotel, which had little to offer in the way of food and drink except slices of bread and butter and soft drinks. As this was better than nothing, we finished our puritans' supper and retired to our rooms.

The BBC was ready to start filming, which commenced in the market place in central Bertoua. John wore what would become his trademark soccer shorts. The filming took up most of the morning, with the cameras rolling as we busied ourselves by purchasing food supplies and other small items for the expedition. Pierre continued to experience some difficulty in acquiring a suitable vehicle for the BBC, which necessitated another overnight stay in Bertoua. The following morning, we joined Jo, Warren, and Nigel at their equally non-descript hotel for lunch, while Pierre negotiated the use of a red Toyota pickup for the next step of our journey to Yokadouma, seven hours south. As we pressed on along the unspeakable dirt road, I decided to take a nap on the journey and let John, Scott, and Rob admire the scenery, which was little more than a sea of green on either side of the track. The journey was slow and punishing, but our safari bus proved to be remarkably reliable. Two hours later we pulled into the Montagna Hotel for cold drinks, before taking lunch at the Crocodile Restaurant. As our day progressed into early evening, we would not be able to reach Youkaduma before darkness fell, and pressing on in the dark was not an option on the dangerous, unpaved road.

Rather than seek out yet another bush hotel, Pierre directed our driver to pull in at a neat, well-kept camp with several single story buildings. The "camp" turned out to be a convent which was run by the Missionaries of Mercy, the same religious order founded by Mother Theresa. We were greeted by a tall young Frenchwoman in her 20s, who could have easily passed for a model and would not have looked out of place on a catwalk in a Paris fashion show. Her modern clothes and short, stylishly cut hair were completely at odds with the regalia normally worn by nuns, even in the tropics. Only the crucifix around her neck hinted that she possessed any kind of religious faith. She quickly arranged for three very comfortable rooms and overnight parking in the safety of the compound.

On our resumed journey the next day, Pierre suggested that we stop at the town of Batouri for a bathroom break and to purchase some refreshments, to which we readily agreed after traveling for three hours nonstop. After alighting from our vehicle, a young Cameroonian man sitting nearby turned to his friend and said, "These must be the people looking for the dinosaur." John and I were somewhat taken aback by this. How did he know who we were and what we were doing in the country? We had deliberately kept our expedition quiet and avoided all publicity. Yet a young African living in a non-descript one-horse town seemed to know who we were. Another problem arose when the BBC crew, still an hour behind us, lost a wheel on the road and were later held up in Batouri with an air conditioning problem. In Africa, one is far better off learning to live without air conditioning, as it is essential to become acclimatized to the heat if one is to work efficiently in either the cloying heat of the dry season, or the cool climate of the wet season. Africa is no place for anyone who enjoys all four seasons of the year!

We later learned that a BBC film crew had visited Moloundou some weeks prior to our arrival in Cameroon. Although the BBC/Discovery Channel filmed all over Africa, including Cameroon, which was famous for its stunning variety of bird life, why would they go to Moloundou? Perhaps they were hoping to "capture" a *la'kila-bembe* on camera or at least interview eyewitnesses for the record.

After another two hours on the road, we stopped at another small dusty town where the chief of police checked our passports and government papers. Although John and I held joint British and Canadian citizenship, we both chose to travel on our British passports, which had "European Union" added to the cover. This seemed to impress our friendly *gendarme*, who wanted to buy us some cold drinks, but was ashamed of the fact that his town didn't have any electricity! No one seemed to know where Pierre and the BBC crew were, so we decided to wait an extra hour in the blistering heat to see if they would catch up with us. After waiting more than an hour, we speculated that the BBC team was ahead of us and may have by now reached Yokadouma. We pressed on and reached the town 90 minutes later. After booking in at the Elephant Hotel, we unloaded our safari wagon and stored all our equipment in a secure brick outhouse. Pierre showed up an hour later, telling us that his vehicle had run into "problems," then disappeared again without further explanation.

Thirty minutes later, Jo arrived at the hotel and stormed into my room barely moments after I had stepped out of the shower with a towel wrapped around my waist. The BBC's vehicle had lost a wheel on the journey, its brakes were also defective, and they narrowly avoided being smashed to bits by a logging truck. She further berated me for "doing nothing about it," and demanded to know why I wasn't calling lots of people on our satellite phone. She then stormed out of my room after I answered her in a less than gentlemanly manner. Communication in central Cameroon was very poor, and needless to say, we had no idea that the BBC vehicle was disabled as Pierre did not elaborate on the situation when he briefly visited the hotel. Furthermore, just who those "lots of people" were that I should be telephoning from the middle of Africa, Jo didn't say.

Later that day John, Jo, and I visited the World Wildlife Fund compound in town and met with a helpful Dutch employee named Hank. A very good Toyota SUV resplendent in WWF livery was available for hire with a driver, and would take the BBC team on the final leg of our journey, 85 kilometers to the Moloundou and the *la'kila-bembe* encounter site. After shelling out more money, we were set to complete our journey the following morning. We briefly stopped at the pygmy camp of Welele, and picked up Timbo Robert and three other Baka trackers who would be our forest guides for the trip.

Arriving in Molundou at 2:00 a.m., we stopped at a bar and purchased large, refreshing bottles of Fanta orange drink. John noticed that our Baka guides were sitting outside the bar in a separate group. Pierre explained that pygmies were heavily discriminated against and were often treated very badly by the different Bantu tribes. A sort of tribal apartheid existed in Africa, where one tribe would discriminate, often in shocking ways, against other tribes, especially the forest-dwelling pygmy groups. It is generally believed that that the Pygmies are the direct descendents of "stone-age" hunter-gatherer peoples of the central African rainforest, who were partially absorbed or displaced by later immigration of agricultural (Bantu) peoples, adopting their Central Sudanic, Adamawa-Ubangian, and Bantu languages. Although this view has no archaeological support, there is some common botanical and honey-collecting vocabulary between the Aka and Baka tribes, which are both western

Pygmy populations but speak quite different languages. Although precise numbers are difficult to determine, it is thought that the number of true (indigenous) Pygmies living in Africa today may number as few as 250,000.

After numerous delays, mishaps, and one near traffic fatality, we finally made it to our target. I pointed to the place in the river where a *la'kila-bembe* was observed a year before to Rob, who felt that he was standing on hallowed ground as he stared out into the still swollen Boumba River. Just getting to the target area proved to be no mean feat; it was an adventure in itself. But we arrived and wanted to establish our base camp before exploring the river.

Just as we were unloading our safari bus, a rainstorm suddenly hit us, literally out of the blue, forcing us to abandon our campsite and beat a hasty retreat back into town. Feeling tired, cold, and thoroughly dejected, we booked into a small, dingy single-story concrete building that offered rooms with bathroom "facilities." The rooms were bare except for a few items of furniture and a double bed. The place was alive with cockroaches, which encouraged us not to visit the bathroom at night!

We showered, changed into fresh clothes and met the BBC for dinner at a small sparse restaurant that had little to offer except rice, vegetables, and river fish. After discussing it, we designated the building the "Moloundou Hilton" and vowed to sleep in our tents outside the building if we were ever stranded there again! Our overnight stay at "the Hilton" passed uneventfully, and we were grateful for the first rays of sunshine that broke through the old ragged curtain that stretched across our window.

The day was bright and sunny as we busied ourselves by checking through equipment and supplies, and pumping clean, safe drinking water into 4-liter plastic containers from a freshwater tap using our Katadyn water filter. Warren kept busy by filming the morning preparations while Jo interviewed John about how the trip was going so far.

Arriving back at our campsite on the banks of the Boumba, our Baka guides cleared the camp area with their machetes remarkably quickly. We were filmed putting up our tents, emptying out our large cartons, and generally arranging our camp to make it as organized and as comfortable as possible.

Jo, being the sole female member of the expedition, had her own tent, while Nigel and Warren shared another. The rest of the team shared a very large, 12-person domed tent, which was big enough to store supplies and sleep all five of us, including Pierre. Our Baka guides quickly rigged up a bamboo platform on stilts that served as a "table" to keep our cooking pots, plates, mugs, and other utensils. Two hours after arriving at our camping spot, we were set for the remainder of the expedition.

Later that afternoon, on a hunch, Rob mentioned the name "*mokele-mbembe*" to two young Cameroonian boys who were bathing in the river. One of the youngsters mentioned that his father and uncle were very familiar with the animal, and accompanied Rob back to our camp to tell us more. Soon, a large group of locals had gathered at our camp, curious about the reason we were there. John took charge of the initial investigative questioning, and used our binder to determine who was and was not familiar with the mystery animals of the region. It seemed that those who spent a lot of time on the river, either tending to their fishing nets or traveling to and from the town with canoes laden with produce from their plantations, were most familiar with the *la'kila-bembe*. Testimonials from firsthand eyewitnesses were most impressive and confirmed all that we had heard before. It was an animal that lived in the river; it sometimes had strong, rigid spikes running the entire length of the head, neck, back, and tail. It spent a lot of time eating—mainly leaves and fruits. It could stay under the water for a long time. It would sometimes surface under a canoe and break the vessel in two. It was dangerous when approached, and would battle elephants and crocodiles. *La'kila-bembe* always won these battles—nothing could ever defeat it! As fortune would have it, the father and uncle of the young informant walked into our camp and related an amazing story. These brothers, Samuel and Phillip, related how, just two years before, they were paddling slowly down the Ngoko River after sundown when they struck a large solid object almost in mid-river. Puzzled by this unexpected collision, Samuel raised his kerosene lantern to view the nature of the obstruction. The object suddenly seemed to come alive, shifting slightly in the river, revealing its identity to the Samuel and Phillip. As their eyes became accustomed to the darkness, the brothers could see by the dim glow of the lantern that the object they had struck in the river was nothing less than an enormous living creature! The body was gigantic, as big as any elephant that they had ever seen. Its neck was characteristically long and slender, ending in a small, python-like head. Before they could react to this sudden encounter, the monster slowly turned its head and stared directly at the two men. The yellowish glowing eyes fixed on the fishermen as they could only stare back at the behemoth. The two brothers immediately backed away from the monster, paddling as hard as they could in the opposite direction, while the animal continued to browse on the tree branches overhanging the river, seemingly unconcerned about the human intruders. That same night, in spite of their encounter with the river monster, the two men tried to continue their journey, but gave up after two more attempts. The *la'kila-bembe* was still there after several hours, confirming much of the information we had collected concerning the animal's habit of browsing for lengthy periods of time along the river's edge. It seemed that the animal was very well known in this region, which prompted us to speculate that Boumba, Ngoko, and Dja rivers may well be migration routes that allow the animal to move freely between Cameroon and the Congo Republic, where the Sangha, Likouala-auxe-Herbes and Bai rivers further serve to facilitate these far-ranging animals as they seek out a plentiful food supply well away from other large animals, river traffic, and the fishing activities of humans in their canoes.

Those locals who spent a good deal of time on the river had observed a *la'kila-bembe* on numerous occasions at the confluence of the Ngoko and Sangha rivers, browsing on the leaves of one particular tree, located on the Congo side of the river. The Lingala name for the animals, *mokele-mbembe*, was mentioned several times, due to the local river traders and fishermen who had observed the animals from both sides of the river. As we continued to question locals about other mystery animals, two more illustrations provoked positive responses.

One of these was a drawing of a strange bipedal ape called the *kalanoro*, drawn by Harry Trumbore and featured in *The Field Guide to Bigfoot, Yeti, and Mystery Primates Worldwide*, by Loren Coleman and Patrick Huyghe. Fearful mutterings arose from the crowd as the name "*dodu*" escaped from the lips of several people. Although Timbo and other pygmies had identified the creature for us the year before, it was gratifying to see others who were equally as familiar with the mystery ape. One young Cameroonian, who spoke very good English, recalled how just a few months before our arrival, a hunting party consisting of several white Europeans and their native trackers actually arrived in Moloundou with a captured *dodu*! The animal was firmly bound to a strong wooden pole by its hands and feet. The description was consistent with the previous independent eyewitness accounts. The creature was grey all over, with a vaguely chimp-like face, three digits on each foot, and rows of porcupine-like quills running the length of its back. The strange creature wailed mournfully, sending shivers of fear down the backs of the sizeable group of townspeople who had gathered to behold this fearsome half-man, half-beast from the forest. What became of the animal is not known. The hunting party moved on, taking their trophy with them. Is the *dodu* captive in a logging camp somewhere in Cameroon? Or did its European captors simply release their prize unharmed back into the forest? Whatever the outcome, a major new discovery of an extraordinary bipedal primate may have been lost yet again.

The following morning was spent filming in the forest. The BBC took several "takes" of us marching through the lush green rainforest in single file, as if we were heading into new and unexplored territory. During a break in the filming, we questioned our guides about the *dodu* again. The younger pygmies had not seen the animal before, but the pygmy elders had indeed encountered them on hunting expeditions deep in the emerald forest. The pygmies would rather do battle face-to-face with a leopard than an outraged *dodu*.

Two of the pygmy elders, Rosé and Paul, recalled stories they had heard about this creature. Paul related how his father had actually fought with one many years ago and killed it with his elephant spear. Rosé told us that several of his fellow Baka had seen a *dodu*, but kept their distance, as this was a very dangerous animal which did not hesitate to kill gorillas and chimpanzees whenever it encountered them. Even Pierre Sima, our fearless elephant tracker, had followed the strange three-toed footprints of a *dodu* in the forest for seven days. He and his Baka team came across a large clearing in the forest which had been seemingly been torn up as if a large group of animals had been fighting. The Baka were able to determine that a sizeable group of gorillas had bedded down in the clearing the night before, but a *dodu* had charged into the group, causing them to scatter in alarm. They also speculated that most of the destruction had probably been caused by the *dodu* engaging the alpha male gorilla in combat. As the party continued to track the creature's footprint, they became aware of a strong, pungent smell, which revealed that they were closing in on their prey. As Pierre recalled later, the Baka became afraid and did not want to continue. But next time, he would take his shotgun!

Rosé also explained that although a *dodu* did not devour the gorillas and chimps it had killed, it would use the mutilated carcasses of its prey to attract maggots. Once a *dodu* kills an animal, it tears its abdomen open and then leaves the carcass to fill with maggots. The *dodu* will later return to the carcass and scoop out the maggots to devour them with gusto. Timbo also mentioned that a group of Baka hunters killed a *dodu* in 1983, then sold the corpse to a Frenchman as a curiosity. The Frenchman and his trophy left Cameroon and were not heard of again. I discussed the possibility of conducting a forest expedition with Pierre to try and track a *dodu*. Needless to say, it will be an armed expedition, should we ever mount one!

On the morning of February 28, we visited the town officials, including the mayor, Jean Jacques Ipando, the chief of police, Yves Patrice Ngondi, and another senior bureaucrat known as the *Sous Prefet*. Like most countries in Africa, Cameroon seemed to be full of petty officials, each determined to ensure that our papers were in order and that our passports possessed the correct visas. John once again smoothed the way with his imposing presence and excellent command of the French language, explaining to each official the nature of our mission, and where our camp would be located if they cared to visit us during the course of our expedition. After the formalities were over, which took up most of our morning, we grabbed an early lunch before returning to our camp. The film crew later interviewed each expedition member individually, starting with John, who was seated in front of a spectacular bamboo tree. Pierre's interview followed, then Scott and Rob, with my own turn in front of the camera reserved for another day. As the evening began to encroach upon the cool of the day, a strange light appeared in the sky at around 4:00 p.m., a dull luminous object that appeared to be moving up and down, almost in a "bouncing" motion. John asked Warren Bradshaw, the cameraman, to see if he could zoom in on the object through his camera lens. Warren was able to focus in on the object but he was not exactly sure what it was. After about ten minutes, John spotted a second object, hovering just above the tree line and about the same distance away as the first object, but much brighter in luminosity—moving from north to south, perhaps only a hundred feet above the trees to the west of our camp across the Boumba River at a mile distant. The second object sped towards the first glowing object, made contact and became one glowing orb, and then several seconds later took off into space at a 45-degree angle at a speed that no known earthly craft could possibly match. It was literally gone from sight into the sky in just under two seconds. The velocity at which it moved left us all speechless. None of us could believe what we had just witnessed. However, the excitement was not completely over.

Later in the evening, just as the sun was slowly sinking over the vast expanse of forest that lay before us, I wandered down to the river's edge to join Scott. Every evening, Scott would take his mug of coffee and place his comfortable, folding camp chair at the river's edge, giving him a panoramic view of the northern stretch of the Boumba. It was a tranquil and relaxing part of the day, as the last few local residents finished bathing in the river and fishermen were slowly paddling their canoes home with their catch. As Scott and I were chatting, a bright ball of light suddenly shot across the sky barely 50 feet above the trees, heading northwest at incredible speed. Astonished at this sudden spectacle, Scott and I could only exchange startled glances before joining the others for dinner at the camp.

Shortly after our incredible observation, John brought our attention to yet a fourth strange object traveling across the sky at a height of about 30,000 feet from east to west. It was not traveling as fast as the second "UFO," but it was much faster than any aircraft or satellite, and was certainly not a shooting star or any other known aerial or atmospheric phenomenon. We lost the object when it crossed over the horizon and vanished. We were amazed that we could have seen four apparent UFOs in just one night. Before retiring for the night, we speculated on what the strange lights might have been. They certainly weren't aircraft or helicopters, meteors, or even satellites. Even if one dismisses the usual culprits that are often for mistaken for UFOs, such as weather balloons or even the planet Venus, these do not explain the brilliant ball of light that zoomed across the Boumba perhaps only 30 minutes after our first initial sighting over the dense forest. Putting the matter to rest, we retired early for the night in anticipation of more filming and actual river work the following morning.

The day started early for us and we were up with the sunrise. Warren was up before us and decided to film us crawling out of our tents, brushing our teeth, and preparing breakfast. Pierre had arranged for the rental of a very large 30-foot-long canoe from the *Chef du Marine* in town, and the mayor kindly rented us his large outboard engine for our river research. Scott and John were paired together and sent upriver with the BBC team. The idea was to film them floating downriver while operating the fish-finder/sonar unit that Scott, our computer and electronics whiz, had brought, along with the satellite phone and a GPS unit to record the exact locations of interest along the river. As they took turns operating the unit, John felt that wrapping the transponder wires around his wrist would not be a good idea, and Scott duly agreed. During Scott's turn to operate the sonar, his focus was on the 2-D screen as it revealed small shoals of fish, large single fish, and crocodiles, when suddenly an unseen creature seized the transponder and tore the cables clean away from the main unit in the canoe, rendering the entire unit useless. Although Scott was able to save what was left of the sonar unit from being dragged into the river, he could have been pulled in had he wrapped the cables around his arm.

John and Scott returned to the base camp as Rob and I were preparing our remote game cameras for use along the river. These were simple 35mm cameras placed in protective metal boxes and affixed to tree limbs. The camera would take a photograph if an object crossed the infrared beam, which in this case would be on the river rather than on land. We arranged to place these at two spots on the river where locals had observed *la'kila-bembes* before. Hopefully, they would pick up some photographic evidence for us at places where no humans would disturb the tranquility of the locations.

Later that day we crossed the Boumba on the floating platform (ferry) and visited Langoue, a Baka village located between the Boumba and the Dja. The mood was somber as we arrived in our safari bus, and Pierre asked the chief if we could speak to the village about our mission. He agreed, and the village people and elders gathered before us, about thirty people in all. Pierre explained in the Baka language that we were visiting Cameroon on a safari, and that we were interested in observing and filming the wildlife around us. After Pierre handed me the binder, I began to show the people our animal illustrations. None of the North American animals drew any recognition from the assembled crowd. The well-known African animals were picked out quickly. Almost all the dinosaur illustrations were rejected, except for the sauropods and the *Triceratops*. These, according to the village elders, were familiar to them.

The *Diplodocus* and the *Brachiosaurus* were referred to as *la'kila-bembe*, and the *Triceratops* as the horned killer, the *n'goubou*. At least four witnesses came forward who had encountered the *la'kila-bembe* in the Dja river. These were either fishermen or plantation owners who sold their produce in the villages along the river. Perhaps the most intriguing aspect of these reports is that the eyewitnesses picked out the *Diplodocus* as closely resembling the sauropod configuration of the *la'kila-bembe*, but in some cases the animals also sported the dermal frills or spikes possessed by the *Brachiosaurus* as illustrated in our binder. Nanga Norbert and Philip Ezeze were plantation owners from further north up the Dja who were visiting the area on business during our expedition. They have both observed *la'kila-bembes* in the river over the years, and Nanga mentioned that he and his father encountered one such monster that was so big, it spanned almost the entire length of the river at the point where they tried to paddle past the animal. The monster turned and regarded Nanga and his father for a few moments before returning to browse on the leaves and fruits by the side of the river. The two men beached their canoe and just sat on the opposite bank behind some bushes, watching the behemoth feeding for what they estimated to be at least three hours! This confirmed many other reports that the animals browse for considerable periods of time in the most tranquil parts of the river and swamps.

Further details from our two witnesses revealed physical features that suggest sexual dimorphism in *la'kila-bembes*. Philip stated that the animals he encountered feeding in the river had distinctive dermal spikes running the length of the head, neck, back, and tail, while the larger animal observed by Nanga and his father three years before our arrival did not possess this characteristic. This greatly intrigued us, as dermal spikes were a physical feature of some sauropods unknown to western paleontologists until the 1990s. This gave us another measure of accuracy in establishing zoological features of the *la'kila-bembe*, and perhaps brought us closer to establishing an actual identity for the animals, regardless of how sensational or unlikely that might be.

Perhaps the greatest surprise of the day came when I turned the final page in our illustrations and showed the drawing of the *kalanoro*, recognized by the Baka people as the *dodu*. Almost as one, the entire Baka group erupted with a mingled roar of laughter, gasps of astonishment, and moans of fear. Yes, they knew this animal well and they feared it. They knew of two incidents when Baka hunters had killed *dodus* with their spears. If at all possible, they avoided all contact with the *dodu* due to its ferocious nature. After we finished, we thanked the village and their chief, Daniel, a veritable pocket Hercules, for his very kind assistance. Scott and Warren had successfully filmed the entire interview, and the obvious sincerity of our village hosts would have impressed even the most hardened skeptic. This ancient continent never gives up its secrets easily—especially to inquisitive outsiders.

Before departing from the village, one of the local fishermen mentioned that the Dja was actually a short walk from the village. We took the opportunity to see this mighty river for ourselves, as this would be the focus of our river exploration the following morning. Standing on the banks of the river, the wind had died, and the last yellow rays of the early evening sun cast shimmering fingers of gold upon the glass surface. How many *la'kila-bembes* lurked within the dark depths, ready without warning to vanquish the tranquility of this river of ages? I quietly took a photograph to capture this rare and all too fleeting moment. Would tomorrow reward us with a new discovery?

In addition to the giant blue dugout, another, much smaller boat was made available for our use. This was a 7-foot aluminum dinghy we could use to float quietly downriver. Rob, Pierre, and I would be towed up the Dja and then proceed to float downstream with the current and conduct quiet observation of our surroundings. John and Scott would conduct further long-range observations with their video cameras from the riverbank. Just as we were about to get underway, John noticed a sizeable splash that emanated from the west side of a large island in the main confluence of the Dja about 100 meters north. Pointing to the location, John felt that we should investigate the source of the disturbance immediately to see what could have caused such a large displacement of water. The island was perhaps 400 meters long and 100 meters wide, creating a narrow channel perhaps 15 meters wide on the Congo side of the river. Something about this place made us feel uneasy. Pierre could sense it more keenly than Rob and I, perhaps because he had spent much of his time in the bush, tracking elephants for eco-tourists with the Baka, and was therefore like them more attuned to wild mysteries of nature in this pristine wilderness. The splash John had seen occurred at the southern tip of the island in the main channel of the river and could only have been made by something at least the size of a hippo. The water seemed deeper there, perhaps up to twenty feet, but hippos were not found in that area because of the presence of *la'kila-bembes*. Without the use of our sonar unit to probe the depths, our efforts were futile, so we decided to head much farther upriver and float down on the dinghy.

As Rob, Pierre, and I headed north on the river, a loud splash emanated from the left bank on the Congo side of the river, north of the island we had just passed. Pierre was the first to spot it, but our boatman was unwilling to stop. He was determined to keep going until we were far enough north to allow us to float quietly south with the current. Was the splash another chance encounter near our mysterious island?

As Pierre prepared dinner that evening, yet another group of officials, this time "immigration officers," arrived at our camp to examine our papers and passports yet again. The lack of progress in many African countries and the hesitance of many potential investors to assist with their development did not surprise me in the least given the way that they buried themselves in mountains of complex red tape and slow-motion, corrupt bureaucracy. Jo threw another fit over the lack of variety in our diet, and the fact that we did not have enough food to last for at least two weeks filming in the forest. As we were exploring a relatively remote part of Cameroon, there was only so much food we could transport, and the difficulties and lack of fast, reliable transportation, not to mention the dangerous roads, had put paid to the amount of time we had left for forest exploration. Besides, the animal of interest to us was a semi-aquatic herbivore that spent the vast majority of its time in the rivers and swamps. Jo later apologized to John for her behavior, in spite of the fact that I was the one who always seemed to bear the brunt of her outbursts. Later that evening we went into town, minus the film crew, for cold drinks and to discuss our dealings with them and how to best use the remainder of our time in the field.

On March 3, at 5:30 a.m., I was awakened by my rumbling stomach, which sent me dashing out of the tent and behind the nearest bush. I felt like my insides were on fire as I doubled over in pain. Twenty minutes later I tried to navigate my way back to the tent in the inky darkness without the use of a flashlight. I was convinced that I had left a trail of burning coals behind me as I hastily exited the tent. There was no doubt that I had picked up an intestinal parasite somehow, whether it was something I ate or even drank. I suffered with bouts of stomach pains and diarrhea for the rest of the expedition, which killed my appetite and drained my energy levels considerably. John took over the rest of the expedition in admirable style, using his previous experience as a television producer in Hong Kong to work closely with Jo before we left the area. There was no question that John was the true leader of the expedition from day one. I felt embarrassed even with the thought of comparing myself to him. Just after midnight, a panic broke out in Nigel and Warren's tent as their cries of alarm and rapid exit from their sleeping bags woke up the camp. Pierre dashed into the tent and quickly spotted a small black mamba with his keen, hunter's eye, as it slithered into a corner behind some film equipment. With the use of a stick and fast reflexes, Pierre seized the serpent, snapped its neck and threw the writhing corpse into the forest. The drama was over quickly and we could all get back to sleep, but not until we searched the remaining three tents first!

That morning, the crew and I headed out onto the Boumba and located a large gravel bank where I was interviewed. Later in the day John was filmed retrieving our game cameras from the trees to which they were affixed, and the film crew returned to Langoue after lunch to film the Baka people conducting a special dance in traditional tribal clothing and masks. Rob, Scott, and Pierre had been towed far up the Dja by the motorized canoe to drift south and make whatever observations that might advance our search for *la'kila-bembe*. I spent the rest of the day at the camp, nursing my stomach and keeping watch for unwelcome visitors.

Throughout the expedition, Scott had taken a close interest in the plight of the Baka. The way they were treated by other Africans appalled him. The destruction of their traditional forest home and their forced eviction to completely unsuitable camps on the side of the logging road outraged him. Their continual suffering from sickness and disease broke his heart. Whatever medical supplies we had, he put to good use in attempting to relieve some of the suffering among our Baka guides. Whether it was medicines, items of clothing, or better quality food, Scott was always willing to render assistance.

Scott revealed a tender heart and a willing spirit to help the downtrodden wherever he encountered them, and Africa was no exception. Even before we had departed from Cameroon, Scott was already thinking of ways to provide long-term practical help for the Baka. His determination and passion to break the cycle of discrimination against these gentle forest people was inspiring.

Jo, Nigel, and Warren returned from Langoue, seemingly pleased with the results of their filming there. We were all interviewed for one last time before dinner. Tomorrow we would pack up and leave Moluondou for the return journey north. Christoff, the chauffer from the Meumi Hotel in Yaounde showed up with the Toyota SUV to collect Jo, and the WWF vehicle also arrived on schedule. It was a pity to just pack up and leave, considering that we were in the target area, and Scott was willing to stay on for a few weeks more. But time was against us, and the difficulties and dangers encountered on the journey to the Boumba had taken up more time than we had anticipated. But we had learned many valuable lessons and would be better prepared for the next expedition for certain!

We had made excellent progress and agreed that we should return to Moloundou again as soon as new funding would allow us to do so. But how soon could another expedition be mounted?

After dinner, we relaxed over coffee in order to gather our strength, as we would break camp at sunrise and get on the road as soon as possible. Unfortunately, our evening was rudely interrupted by the arrival of two individuals. They had been drinking and were carrying a shotgun and a machete.

Scott was the first man they encountered. Before we could ask what their business was, they loudly demanded alcohol and money in exchange for two African grey parrots they were carrying. They were illegal traffickers in rare birds, and they were very unpleasant characters indeed. Although he was unarmed, Scott stood tall and fixed the two intruders with his steady stare from under his trademark fedora hat, while John, Rob, and I scrambled to grab our machetes. Surrounding the two poachers, we made it clear that they were outnumbered and should leave immediately before we overpowered them and handed them and the stolen birds over to the police in Moloundou. This was something they did not expect. European loggers might be willing to buy their illicit goods, but we were not to be trifled with. Quietly, they turned around and slunk away into the darkness. Thankfully, they did not return.

By 7:30 a.m. on March 5, we had packed up our camp and loaded up our safari bus. We bade farewell to Jo, Nigel, and Warren, and I made a point of thanking her for the hard work and dedication she and her crew had put in to the expedition. I also apologized for the misunderstandings that arose between us, which certainly were not all her fault. Missing luggage at the airport, lack of suitable transportation, and the various mishaps on the journey certainly added to the frustrations felt on both sides. We hoped to see the film crew in Yaounde. The minute we vacated our campsite, a large group of locals immediately descended on the area, collecting empty water bottles, a few old pots and pans and other odd items we had left behind for them on Pierre's advice. Apart from a brief stop in Moloundou for breakfast, we hit the road, determined to reach Moloundou before the midnight hour. The drive was long, tedious, and exhausting. We stopped only to answer the call of nature, and finally reached Pierre's plantation in Dimako at 9:30 p.m. Phil and Reda Anderton greeted us and prepared a welcome dinner of spaghetti with sauce and cheese.

Pierre went into town and found a decent Toyota minibus to ferry us back to Yaounde. After breakfast we wasted no time in loading up our luggage in the vehicle and proceeding on. The journey was much slower than usual, with the narrow muddy road and rickety bridges being filled up with a long line of very slow heavy traffic moving in the opposite direction. Pierre explained that hundreds, if not thousands, of Muslims were heading over the border to the Central African Republic for a religious gathering. We stopped only once to seek refreshments in a small nondescript town, finally reaching the Cameroon Bible Institute at 1:30 a.m. We were given the use of a rather nice comfortably furnished house, normally used by Walter Loescher and his family, who were in the USA on furlough. John and I grabbed the master bedroom with a ceiling fan and en-suite bathroom while Rob and Scott found a bedroom each.

Our final four days were spent relaxing at the institute, enjoying home cooking, purchasing souvenirs in town, and preparing for the flight home. At $7.00 US per day, which included accommodation and meals, the institute was considerably cheaper than the hotels in Yaounde! There was a team of American volunteers also staying at the institute, involved in construction work, building repairs, and the installation of an internal telephone system. At dinner one evening, one of the volunteers decided to crack a joke referring to Chinese children and "slitty eyes," not knowing that John and I both had Chinese spouses and mixed-race children! We told him later, much to his embarrassment.

On the evening of March 3rd, we flew out of Yaounde and headed for Paris, where Rob and Scott parted company from John and I to catch their flights back to the USA. They had proved themselves to be true adventurers and explorers. No doubt they had been bitten by the explorers bug and had undergone some life-altering experiences that would remain with them for eternity. John and I were back in Canada within 24 hours of leaving Africa. My journey ended in Toronto, while John continued on to Vancouver.

News reached us that the Discovery Channel had decided to shelve the documentary due to insufficient film material. Jo Sarsby never contacted us again, but Greg Richardson did. He was more than disappointed with the outcome of the expedition, referring to us as "amateurs" and critical of my choice of team members. He also mentioned that he was hoping to arrange for a second team to fly out to Cameroon to continue where we had left off. We were all opposed to this, as our research and the information we had acquired had not come easily, not to mention the extreme hazards of the journey just to reach the target area. Greg parted company from us, and never did send out a second team to continue with our research. It was a pity, nevertheless, that he had lost confidence in working with us again. John, Scott, and Rob had gained invaluable experience in the field and therefore knew exactly what would be required of them if another expedition could be organized. We were ahead of the game and knew that we would have to return to Cameroon sooner rather than later.

John and Scott decided to set up a new organization to serve as a focal point for our research in Africa and worldwide. The new organization was called CryptoSafari and was inaugurated with a smart new website designed by Scott. Although I was happy to join the new organization, I felt that it had perhaps moved too far from its Christian/creationist roots. After much thought on this, I decided to bow out of CryptoSafari and establish another organization called Creation Generation. This was not to rival the work of John and Scott, but to compliment and support them. After all, we had been through a lot together and had forged a deep bond of brotherhood. Hopefully we could all be back together in Africa one day for a more extensive and successful expedition. My final task was to send a full expedition report to Paul Rockel. In spite of the fact that we had not observed a *la'kila-bembe*, the expedition had placed us in an invaluable location where the animals were indeed present during our time on the river system. We simply did not know it at the time, but the next expedition was to provide us with far more information on the habits of the animals, and where they could be found for certain!

> That animal I saw,
> it was a Brontosaurus!
>
> —Marcel Bobas

7
ADVENTURES IN THE FORBIDDEN ZONE

The expedition had, as usual, provoked a great deal of interest among creationist organizations around the world, many of which posted our report on various websites. This caught the attention of Brian Sass, a board member of a Saskatchewan Creation Science organization, who was interested in the search for *mokele-mbembe*. He wondered if it would it be possible to participate in an expedition to Lake Tele, and did I have any photos of that location? I emailed a reply to Brian with two attached photos of Lake Tele that I had taken back in 1986. However, the Congo was still recovering from yet another civil war, and our research was focused on Cameroon near the border with the Congo. Brian persisted with his emails, seeking more information, as he was still interested in coming along. Field research in Africa would be a dream come true for him, and he would do his best to raise some funding for the trip.

In early 2002, my family and I had moved to Calgary, Alberta. Terri had more family immigrating to Canada, and had a brother and sister already living there, with other relatives in and around the city. We quickly settled into our new hometown and established ourselves in new jobs. The Rocky Mountains, Banff National Park, and other glories of the West were a huge contrast to the concrete jungle of Toronto and the popular but overcrowded beaches of the Scarborough Bluffs. Alberta was the western province next door to Saskatchewan where Brian Sass lived, and as he was due to visit Calgary in the near future, we arranged to meet in order to discuss the possibility of working together in Africa, with all the risks and complications that such an undertaking involved.

On a pleasant sunny day in April, I was sitting in the office of Ron Swanson, the Vice Dean of Victory Bible College, located on the Trans-Canada Highway just west of the city. As I was running through a slide presentation with Ron, who had invited me to teach a basic creation science course at the college, Brian Sass arrived with his lovely wife, Denise, and their two children, to meet with me. We later convened to a local McDonald's restaurant where we enjoyed lunch and discussed the possibility of putting together another expedition to Cameroon. Like David Woetzel before him, I found Brian to be a solid family man who was an employee in good standing with IBM Canada for 20 years, and someone who possessed a keen scientific mind. I gave him a short history of all previous expeditions, including the trials, hardships, and frustrations that often went with such adventures. Brian approached the subject in a very down-to-earth manner, and was not in the least given to romantic illusions of derring-do adventures in a "lost world" filled with prehistoric monsters, cannibal tribes, and muscle-bound native porters who addressed every white man as "Bwana."

After our meeting, our families parted as friends and Brian returned to Saskatchewan, determined to raise interest, and funding, in our newly proposed expedition. Generating interest in a "dinosaur hunt" in Africa was easy enough (particularly if the tabloids got involved, which was just the sort of publicity that I vigorously shunned), but raising sufficient funding was much more difficult.

By the turn of the 21st century, an enormous amount of misinformation on *mokele-mbembe* was already circulating on the Internet. In an attempt to maintain the accuracy of the history of the search for *mokele-mbembe*, I was invited to write an article on the subject for *Impact*, a glossy color circular published and distributed by the Institute for Creation Research, then based in El Cajon, California. This led to more letters and emails from all over the world from people who wanted to join any future expedition that might be returning to Africa. Daring and romantic though such adventures sounded, the harsh realities of working in remote and primitive areas under very basic conditions would break many a spirit.

While Brian worked on raising awareness and funding among the churches in Saskatchewan, I received an invitation from Dr. Dennis Swift to speak at his 16th Annual Creation Science Conference in Portland, Oregon. Dennis was well known for his own "living dinosaur' research, and was widely regarded as the world's foremost expert on the famous Ica Stones of Peru, and the dinosaur clay figurines of Acambaro, Mexico. A church in Washington State was also interested in having me speak during the conference, which was guaranteed to keep me pretty busy during the three days I was due to spend in Portland. In October 2002, I had completed three different talks at the conference in Portland, and answered what must have been a thousand questions from interested attendees who crowded around my book table. Thankfully, my good friend and superb wildlife artist, Bill Rebsamen, from Fort Smith, AR, was there to lend a hand. Bill had illustrated my first two books, *Claws, Jaws & Dinosaurs*, and *Missionaries and Monsters*. Towards the end of the final night at the conference, I was approached by Milt Marcy, an insurance broker who lived in town. A huge man, with a deep, 1950s style radio announcer's voice, Milt was somewhere in his early 50s and had maintained a strong interest the history of alleged living dinosaurs, including the various expeditions that had gone in search of them. Milt laid it on the line—if he could come along on the next expedition, he would pay for the whole thing. Having had similar offers before that had quickly evaporated, coupled with the fact that my brains had literally turned into rice pudding after three solid days of lecturing, I was barely able to muster enough interest in his offer to reply with a coherent answer. After a brief chat, Milt left and I finished my speaking engagements with a talk at a local church before returning to Canada.

Two day later, I received an email from Milt, expressing his disappointment at my vague response to his desire to help, and that any assistance he was willing to offer in the way of funding had now been rescinded. I replied to Milt's email and explained in detail how the conference had kept me busy to the point where I was completely exhausted at the end, and that any apparent rudeness or disinterest in his offer was certainly not intended. I completed my email with an apology and sent it off.

Brian had started to generate some interest with funding for the expedition, and Milt replied back to my email, in the affirmative, asking me to call him to discuss what assistance he could offer. As soon as 2003 came along, the months began to pass by quickly. I drew up a list of equipment and supplies we would need. Most of our base camp gear was safely stowed away in Cameroon, leaving only clothing, food supplies, gifts for the Moloundou officials, and a video camera for myself. We aimed for an October departure, which would have been the beginning of the wet season. My biggest worry was that Francis Ngwa Diba could not be contacted, and Pierre was unable to find him anywhere in Yaounde. This left us with no direct contact with the Ministry of Scientific Research and Innovation, Madeleine Tchuinté, which would be crucial if we were to acquire the necessary documents that would allow our expedition to proceed. Pierre attempted to meet with the minister, but to no avail. We could acquire our visas from the Cameroon High Commission in Ottawa without any problem, but additional government papers were crucial if we were to travel in the country. I telephoned the high commission and explained the situation. I was assured by the embassy operator that if I sent a fax to the High Commissioner, Philemon Yang, he would respond promptly and assist us in obtaining our papers.

Commissioner Yang's reply never came, leaving us with only our 90-day tourist visas to travel on. During our preparations, I received an email from the Reverend Paul Nkang, senior pastor with the Church for All Nations, based in Yaounde. He had heard of our work in Cameroon and decided to contact me from the USA where he was on a speaking tour. I mentioned our plight to him and he promised to help us when we arrived in Cameroon. Our supplies had now been acquired, including other items sent to us from relatives and friends of the Anderton family. After much searching for reasonably priced airline tickets, Dollar Saver Travel in Overland Park, Kansas, came through with two return tickets from Calgary to Yaounde for under $2,000 US each, and on the much loved Swiss Air. Milt had been extraordinarily generous, paying for our additional cameras, airfares, and ground expenses. Although he was disappointed that he was not coming on this particular trip, it would prove to be to his advantage at a later time.

On October 17, Brian flew into Calgary and took a cab to Victory Bible College, where I was teaching a one-day creation seminar. After completing the seminar, we drove to my home to drop off his luggage, then continued on to Wal-Mart to purchase a few small items before our departure the following morning. As usual, I found it difficult to sleep before leaving for an expedition, and constantly went over my notes to ensure that everything was in order. Passport, airline tickets, inoculation card, equipment list, and expedition manifesto. Brian had prepared a binder with high-resolution satellite photos of the Dja and Boumba Rivers, courtesy of Dr. Ed Holroyd, an atmospheric scientist and expert in the field of remote sensing and image processing for the GIS department at the University of Denver. Ed was keen to assist us in any way he could and provided us with over one million dollars worth of satellite imagery of the river and swamp system of southern Cameroon and the northern Congo Republic. This generous contribution proved to be invaluable in assisting us to plot our course on the river system, and to locate and pinpoint areas worthy of future investigation.

Lining up to book in our luggage at Calgary International Airport gave Brian and me some cause for concern. The security staff was searching all the suitcases before they were tagged and loaded onto the plane prior to take off. An elderly lady in front of us had knitting needles removed from her luggage, which concerned us as we had hunting knives and machetes packed away in our own luggage!

As the security guard began to search our luggage, he asked us where we were going. "Cameroon, West Africa," I replied. Suddenly, the man's face glazed over and he said almost dreamily, "Oh, yes, I knew that, please proceed, have a good trip." I looked at Brian in bewilderment and commented, "What the heck was that? I never mentioned where we were going before now." Brian gave a knowing smile as we made our way to the departure lounge. As we sat in our United Airlines Airbus A320, the captain announced that there was a minor technical problem with the plane, and that all electrical systems, including the air conditioning would have to be shut down for about five minutes. In other words, the pilots had a problem that could only be solved by literally rebooting the plane!

Integral to the A320 is the advanced electronic flight-deck, with six fully integrated EFIS color displays and innovative side-stick controllers rather than conventional control columns. In 1988, Airbus was the first aircraft manufacturer to offer airliners with this new computerized digital fly-by-wire flight control system, where the pilot controls flight surfaces through the use of electronic signals rather than mechanically operated pulleys and hydraulic systems. Today, almost every aspect of an airliner's flight control system, including fuel consumption, is computerized.

According to our tickets, the aircraft that would fly us to Zurich was a Swiss Air McDonnell Douglas MD-11, the same type that crashed off the coast of Nova Scotia on September 2, 1998. Faulty wiring started a cockpit fire that rendered both pilots unconscious before they could dump enough fuel to facilitate an emergency landing in Newfoundland. I also noted with consternation that the second aircraft that would fly us to Africa was the Airbus A330. An American Airlines A330 crashed shortly after take off from JFK International Airport on November 12, 2001. The co-pilot flew into the turbulent wake of a previous airliner heading on the same course and used the rudder on the vertical stabilizer (tail) rather than the wing flaps to try and keep the aircraft level. The over-stressed tail completely broke away as a result, causing the plane to crash into Belle Harbor, Queens. I tried to put these distressing thoughts out of my mind as we sped down the runway for take off.

As we approached Chicago O'Hare International Airport, the captain allowed the passengers to listen in on the tower where the flight controllers were guiding airliners in on their landings and take-offs.

One particular controller sounded remarkably like Milt Marcy, guiding in three or four different international flights with his calm, deep bass voice. I speculated that Milt, dissatisfied with selling insurance to people from his offices Monday to Friday, secretly moonlighted in the more exciting field of air traffic control just to keep his voice in shape. Later, as Brian and I walked into the airport, a huge, bronze statue of a *Brachiosaurus* skeleton loomed over us. We hoped it was a good sign.

Three hours later we were seated in our Swiss Air flight to Zurich. I found it odd that since the dreadful events of September 11, 2001, the security staff would remove certain items from our carry-on luggage such as a small pair of first-aid scissors and nail clippers. Later, as the cabin crew was serving dinner, they handed us stainless steel cutlery, including a sharp, six-inch knife to carve up our steaks!

The flight and the service were flawless. It was hard to believe that we were seated in coach, as the quality of service and food were very impressive, and certainly better than anything we had experienced on Air France.

Zurich International Airport proved to be just as expensive as the rest of Switzerland, as a microscopic cup of coffee and an equally diminutive croissant lightened my wallet to the tune of three US dollars. I received change in Swiss francs, which wouldn't be of much use in Africa. The final leg of our flight landed in Malabo, Equatorial Guinea, before finally landing in Yaounde. Pierre was there to meet us and to help us to get through customs. The Cameroonian customs officials would search the suitcases of new arrivals before they left the airport, often confiscating various goods being brought into the country. Cameroonians returning home from Europe often brought computers, television sets, and radios with them, and would have to pay heavy customs "fees" if they wanted to keep their goods. Just as we approached the customs officials, Pierre interjected and spoke with them briefly in rapid French. Without further ado, we were waved through and left the airport without our three suitcases and bulky backpack being searched. This was the second time that we were able to literally breeze through strict customs searches without delay. A French couple immediately behind us was not as fortunate, as their suitcases were thoroughly searched and some electronic items removed, much to their dismay. Pierre also arranged for a courtesy vehicle from the Mercure Hotel to take us into town. The Mercure was more modern than the Hotel Meumi that I had stayed in two years before, with better security, better food, and more comfortable rooms. The staff all spoke excellent English and we were able to pay for our rooms in CFA, US dollars, or Euros.

Our room was pleasantly furnished and we slept from 9:30 p.m. to 4:00 a.m. the following morning. Pierre joined us at 8:00 a.m. for an excellent buffet breakfast. However, his news was not good. He was still not able to visit with the government minister for science, and the only option open to us was a bribe of some kind to secure our government papers. This was not what we wanted to hear, as Brian and I were not willing to bribe our way down to Moloundou. European companies and even missionaries who wanted to work in Africa were often pressed upon by government officials, even at the highest level, to pay certain "fees" (bribes) if they wanted to work anywhere in the continent. No wonder Africa is the only continent in the world going backwards. After lunch, we continued to keep a positive outlook and went ahead with our purchases in Yaounde for the trip south. We decided that the lower Boumba and Dja regions would be worth another search. There was something about that island from 2001 that had greatly intrigued us, but would we be able to reach that location without government papers?

After dinner, as I fumbled with the key to our hotel room door, the telephone rang. I was greeted by the familiar voice of Pastor Paul Nkeng, who had just landed in Yaounde and was on his way to our hotel!

Somewhere in his 30s, Pastor Nkeng was tall and broad shouldered, wearing a dark two-piece business suit. Although he once enjoyed a promising career as a telecommunications executive, he felt that the most productive years would be much better used as a full-time church pastor. Unfazed by our dilemma regarding the lack of official documents that were required to make our trip possible, he immediately decided to make Brian and me "missionary pastors" with his church! We were to report to his ministry office in the morning with two passport-sized photographs and he would take care of the rest.

After breakfast at the hotel, events started moving much faster. Paul Nkeng drove us to his office where Brian and I handed over two color passport-sized photographs each before heading off to the Western Union to draw our expedition funds, sent on by Milt Marcy. Pierre had already arranged our transportation to Moloundou, and the remainder of the day was spent purchasing expedition supplies, including the all-important kerosene cooker. Pierre joined us for dinner at the hotel and would pick us up the following morning. We were concerned that our ministry papers and other documents had not yet materialized, but Brian felt supremely confident that everything was going to work out. I hoped he was right.

Shortly after we finished an enjoyable buffet breakfast, Pastor Guy Bayong, an associate of Paul Nkeng, arrived at the hotel and presented Brian and me with two colorful identification cards bearing our names, dates of birth, nationalities, and titles. We were officially "Missionary Pastors" with the Living Word Church International. An accompanying document, stamped and signed by Pastor Nkeng, stated that we were on official church business. On October 14, with our new credentials in hand, we were ready to leave town. Pierre arrived in the customary Toyota minibus and we were on the road by noon. The journey to Bertoua was surprisingly smooth, and we arrived at the Anderton's home at 5:30 p.m., just in time to freshen up for dinner and prepare for a restful nights sleep. Unfortunately, a heavy rainstorm kept us awake for hours, reminding us that we were truly back in Africa! After breakfast, Pierre heard that the mayor of Yokadouma was in town and suggested that we visit him. Pierre worked tirelessly at establishing good contacts at all levels of society in order to better facilitate our work. The mayor was, unfortunately, ill, and was not receiving visitors, so we continued with our plans and prepared for the final leg of our journey to Moloundou the following morning.

Before leaving town, we picked up Pastor Joseph Nini, who led a small church in Bertoua as part of Paul Nkeng's congregation. Pastor Nini was coming along to assess the spiritual needs of the Baka people in the south, and to determine if and when a church would be required there. We were hoping to hire a Toyota Land Cruiser, but none were available, so we had to make do with a lumbering safari bus from *Alliance Voyage*. The buses made up for their lack of comfort by being remarkably reliable.

As we continued on our journey, it seemed that we were stopping at more and more military or police checkpoints than before. For some reason, we were allowed to pass far quicker with ministry credentials than before with government papers. At one checkpoint, the army personnel just sat and stared ahead, apparently bored, and apparently not aware that we were waiting for them to walk over to our vehicle, a mere ten feet away. After a few minutes we pulled away and continued our journey. The soldiers did not budge from their seats. I felt odd, as if we were somehow invisible.

The road was in its usual catastrophic condition, a narrow river of reddish-brown mud that snaked through the forest and savannah into the distance. I lost count of the number of times that we became bogged down. We were soon caked in mud as we pushed, pulled, and rocked our vehicle out of deep muddy ruts and menacing syrupy pools that seemed determined to suck our two-ton bus into a murky void. Finally, we made it to Batouri where we alighted to stretch our legs, buy some refreshments, and acquire a foam mattress for Pastor Nini. On our way again, our driver decided to make a detour through a narrow jungle road where the low hanging tree branches brutally whipped our vehicle as we sped along. One hour later we were crossing the Kadia River on a floating pontoon, similar to the one employed on the Boumba River. By 5:30 p.m., we made it to Youkadoma and the Elephant Hotel. The rooms cost 10,000cfa per night, in spite of the fact that there was no electricity. The owner of the hotel had neglected to pay the electricity bill, so the hotel employed the use of a generator that provided power from 6:00 p.m. to midnight. Still, a cold shower followed by a change of clothes and dinner of chicken and fried plantains certainly put us in a better mood.

The sun was already high in the bright morning sky as we scoffed down bread, omelets, and hot coffee before heading back onto the mud bath that led to Welele, the pygmy settlement. As time was still not on our side, we had only enough time at Welele to pick up Timbo Robert and his three companions.

By 5:30 p.m. we finally reached Moluondou and visited the town dignitaries to dispense our gifts of T-shirts, sneakers, and hunting knives. Everyone was happy to see us, and the *Sous Prefet* promised to deal with any problems we might encounter in town with the perennial red tape. The new chief of police was less than delighted to see us, and was unsatisfied with his gifts. He immediately demanded a financial "gift," which we were unwilling to pay. However, we did not possess any official government documentation and our church documents did not guarantee freedom of passage outright.

Pastor Nini was most unhappy with the demand of money from the chief of police, and warned him that he was hindering God's work. Standing in from of the chief, the indignant pastor made it clear that as we were "pastors in the land," he would be struck down if he didn't get out of our faces. "God will not forget this day, and you will pay for it!" warned our Cameroonian Saint Paul. Brian and I just wanted to get on with our expedition and reluctantly gave the chief 10,000cfa. In spite of the money, the chief warned that he would be keeping tabs on us during our stay in the area.

Darkness began to fall as we made our way to a small family restaurant for a dinner of fish, vegetables, and rice. Crossing the Boumba was out of the question in the dark, and after dinner we made our way to the now infamous Moloundou Hilton for the night. Once the safari bus was unloaded, the driver headed off into the night and would not return until we were ready to leave the area. We elected to sleep with the light on to keep the mutant cockroaches at bay.

Pierre was up at daybreak and encouraged us to prepare for a quick breakfast before we crossed the Boumba. He returned at mid-day with the usual Toyota mini-bus that would take us to the river ferry with all our equipment. Our 2001 campsite was completely submerged under the swollen river and useless to us. We decided to cross the river to find an alternative site in a quiet location away from Moloundou's overbearing chief of police. The floating pontoon was not in operation, which forced us to cross the river in two small canoes that threatened to capsize against the slow but strong current. An enterprising young Cameroonian showed up at the other bank with a small Toyota car and took us to our campsite in two trips. We arrived at Langoue, a mixed Bantu and Baka village, where the people had given us valuable information two years before on *mokele-mbembe* and the *dodu*. Most of the villagers were away tending to their plantations, and the few women and children that remained turned out to greet us. Timbo and his group quickly cleared an area under a shady tree for our tents, and our camp was soon established. Pastor Nini wasted no time in calling the few villagers to attend an open air service, including songs of praise, prayers, laying on of hands, and a blessing for all who attended. We were grateful to retire for the night after a dinner of fried plantains and sardines. I longed for a cheeseburger and fries! The nocturnal orchestra of tiny insects filled the night, broken only by the blood-curdling shriek of a tree hyrax hidden in the darkness of the nearby forest. Brian had brought along a couple of air mattresses that were used by the Canadian army. These were certainly more comfortable than the thin foam mattresses we were accustomed to using. Sleep was again a brief affair when Pastor Nini began to sing his praises to the Lord at 6:00 a.m.! We had little choice but to get up with the sun, have breakfast, and make our preparations for the day.

The few villagers that trickled in from the bush recognized Pierre and me from our visit with the Discovery Channel film crew two years previously, and still remembered our interest in seeking the *la'kila-bembe*. The day passed quietly as we tested our equipment, explored the immediate area surrounding the village, and talked to some of the Baka about other strange or unknown animals that they were familiar with. One of the stranger creatures I learned about was a gigantic forest spider, somewhat similar in appearance to a tarantula.

One of my correspondents, Ms. Margaret Lloyd from Potters Bar, England, and formerly of Rhodesia (now Zimbabwe), related a curious tale concerning an encounter between her parents and a gigantic spider in Africa. In 1938, Mr. Reginald Kenneth Lloyd and his new bride, Marguerite, were on a motorized exploration of the Belgian Congo. As they were driving along a narrow forest path, they spotted what appeared to be a large jungle cat or a large monkey on all fours, crossing the path ahead of them. As the Lloyds approached the creature, they were horrified to see that the animal was in fact a gigantic brown spider, at least four or five feet in length! Reginald reached for his camera but was trembling so much he was unable to snap a steady photo. Marguerite screamed in terror at the sight of the nightmarish creature, which quickly scuttled away into the forest after briefly stopping at the approach of the Lloyds' old Ford truck. Later, the Lloyds recalled their story to friends, stating that the creature they observed was definitely a spider, reddish brown in color and similar in appearance to a tarantula, but of gigantic size, spanning close to five feet. I asked the Baka if they knew of any such giant arachnid, and indeed they did! They called the monsters *jba fofi*. ("Jba" means "great," and "Fofi" is the general Baka word for any spider).

The *jba fofi* is a spider of similar size to the creature observed by the Lloyds. It makes a ground lair out of leaves, similar in appearance to a traditional pygmy hut. The giant arachnid spins a web between two trees, which is used to ensnare birds and small forest game. One of our informants, Bruno Abolo, drew a life-sized web on the ground and stated that the spider was big enough to cover the entire web, which was very strong. The Baka did not like the spiders and would kill them if they made their lairs too close to their villages, as the monsters were strong enough to overpower and kill a human being. They laid large white eggs about the size of a peanut, producing yellowish-colored hatchlings with purple abdomens. The spiders later changed into a dark brown color as they matured.

Timbo Robert then explained that had I asked him about the spider when he had first met David Woetzel and me in 2000, he could have shown us just such a monster that had taken up residence in the forest behind his camp at Welele! At that moment I felt like drowning myself in the river. A golden opportunity to capture a rare and completely unclassified species of giant arachnid had eluded us. I learned very quickly that the Baka only volunteer information when asked, and one had to be very specific when presenting questions to them about the flora and fauna of the forest.

For a while we pondered on the question of giant spiders. What could they be? As the sun began to sink beneath the equator, we dined on the usual fare of rice, vegetables, and canned meat, washed down with hot tea. Late into the evening, Brian and I stood in the village square and marveled at the incredible night sky filled with countless brilliant stars, like precious jewels cast upon a velvet blanket, crowned by the waxing moon, an amber chariot on its journey through the zenith of the heavens.

Eventually we turned and strolled back to our tent, as a meteor shower streaked across the majestic heavens. The spectacular African sky reminded me of Psalms 19:1: "The heavens declare the glory of the Lord; the skies proclaim the work of His hands."

The following morning we got our first look at the Dja. The water level was high, which made it easier for navigation, and for a large semi-aquatic animal to hide. Brian saw his first really big jungle tree, while the villagers were still strangely subdued. Pastor Nini believed this to be spiritual, rather than just the presence of white visitors. The villagers were still curious about our visit and we took the opportunity to interview them on the many strange animals that inhabited the surrounding forest and rivers. Instead of using our control test illustrations, we first asked the villagers to draw a picture of the *la'kila-bembe* on the ground. At least three Baka each drew passable pictures of the animals, depicting both male and female of the species. We then asked about the *n'goubou*, which they again drew on the ground. The drawings of the river *n'goubou* looked like a sort of hippo with one, but sometimes two horns. The savannah *n'goubou* looked distinctly like a ceratopsian dinosaur, not unlike a *Triceratops* or a *Styracosaurus*. Finally we asked about the *dodu*. At least three witnesses came forward and drew an ape-like creature with a powerful body, long ape-like arms and long porcupine quills protruding from the back of its head and down its back. They feared the *dodu* mainly because of its ill-tempered, aggressive nature, including its propensity for fighting gorillas. Pierre sensed that our questioning should come to a close. We could continue tomorrow. Later, Brian and I explored the nearby forest and discovered a small path that led down to a delightfully shady stream where we could relax in the cool breeze, a welcome escape from the heat of the day. We took photographs of the Baka drawings and caught up with our notes. The village children beat drums and sang into the night, practicing for a forthcoming festival. Sitting in a camp chair with my mug of tea in hand, I relaxed and enjoyed the singing until fatigue finally drove me to my bed.

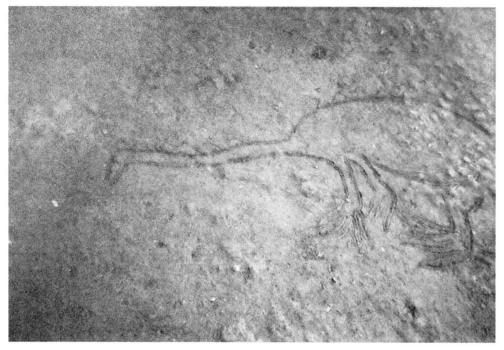

Eyewitness Sketch of a Mokele-mbembe

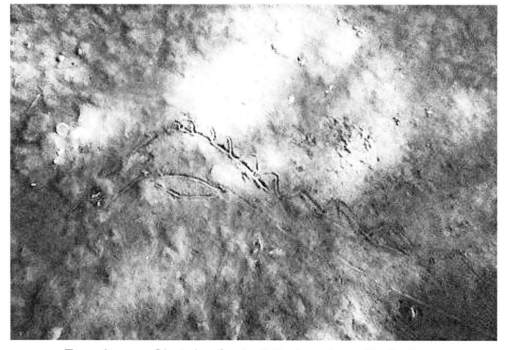

Eyewitness Sketch of a La'kila-bembe with Spikes

Although Pastor Nini woke us up with his praise singing at 6:00 a.m., Pierre was already preparing our breakfast of bacon, eggs, and coffee. Before we could head off to the river, the *Chef de Politique*, Joel Nanga, visited our camp to check our documents. He was the cousin of Norbert Nanga, an important *mokele-mbembe* eyewitness and plantation owner who lived farther north on the Dja. Today we would spend the day on the Dja. A native canoe was waiting for our first day on the river with two young village men ready for the strenuous paddle upstream. As we silently headed north, the surrounding forest canopy was alive with birds and monkeys, and small colorful lizards sunned themselves on the overhanging branches in the water. Pierre spotted a huge Nile monitor, easily two meters (seven feet) in length, but the wary varanid scrambled into the forest before we could even raise our cameras. Thirty minutes later we arrived at the plantation of Norbert Nanga, who happily received us and gave us a tour. During our walk through the coca plants, Norbert showed us a very large spider with a leg span of around nine inches. The intimidating-looking arachnid was black all over with red striping and was highly venomous, according to our host. We left Norbert and the spider behind as we continued on our slow paddle up the Dja. As the light of the day began to fade, we drifted slowly south until we reached our debarkation point and plodded slowly back to our camp. I later tried unsuccessfully to capture the night sounds of the jungle around us, but the steady rhythm of the village drums scotched my plans for the night. We decided to start again early the following morning and bedded down shortly after dinner. We hoped that yet another local bureaucrat would not show up and delay us from getting an early start on the river the following morning. It was imperative that we spend as much time on the river as possible in order to maximize our time.

As usual, Pierre was up early and preparing breakfast. There was nothing like the smell of bacon, eggs, and hot coffee to rouse one from a tent in the jungle! Just as we were preparing to head out, a local man by the name of Monguimbi appeared in the camp and offered to share his own encounter. In late 1987, he was fishing in the Ubangi River in the Congo when a huge animal suddenly broke the surface of the river and moved towards the far bank. After briefly surveying its surroundings, the animal began to forage on the leaves and fruits on the overhanging tree branches at the river's edge. Fascinated by this strange creature, Monguimbi stopped his fishing activities and continued to watch the strange animal as it browsed along the riverbank. He speculated that as he was a good twenty yards away from the animal, it didn't notice him as it began to feed. The river was perfectly calm with no breeze, and there were no canoes in the immediate vicinity. Eventually it grew dark and Monguimbi decided to paddle away to his temporary camp and leave the monster behind. As Pierre began to show our witness pictures of living and extinct animals,

Monguimbi immediately picked out a picture of the *Diplodocus*. "Brontosaurus," he said without hesitation. Brian and I looked at one another. A brontosaurus? "Why did he call it that?" I wondered aloud. As it happens, a few years after the sighting our witness saw a television show on dinosaurs in one of the towns, and likened the animal he saw to the brontosaurus (now called *Apatosaurus*). After completing our notes on this sighting, we thanked our informant and made our way to the river.

As we headed south on the Dja, our guides, Sando and Boima, mentioned that there was another island to the south that was almost divided by a small stream where "big animals" would move around at night. We headed straight for the island, which almost divided the Dja in two. As we slowly paddled through the center of the twenty-foot-wide stream, we could hear something quite large trying to break through the foliage on the island to the left (east of us). Before we could ask our guides to slow down so we could investigate further, Sando and Boima began to chatter loudly to one another, broadcasting the sound of their voices across the tranquil water and through the bush. Two seconds later a huge crash followed by solid splash made by something sizeable entering the river reverberated around us, panicking our guides who headed straight for the west bank. Clearly shaken by the sudden entrance into the river by a very large animal, our guides were in no mood to get back into the canoe. Nevertheless, Pierre pressed them relentlessly on the noise. Was it a hippo or perhaps an elephant? "No, no!" both men stated in union. "Those animals are not found here!" They did not want to discuss the matter further and were unwilling to return to the river. Disappointed at not being able to investigate further, we followed a long winding path through the island until we came to two modern, brick-built houses, complete with wooden floors and comfortable furnishings! The property was inhabited part of the year by French loggers who were surveying the area for suitable hard wood timber to harvest for European markets. During the wet season a local Baka village would act as caretakers until the French returned. The chief, Ondou, was the brother of the deceased chief of Langoue, where we were camped. Like most Baka we had met, Ondou proved to be friendly and accommodating. We asked him about the *la'kila-bembe*; had he ever seen one? Yes, he confirmed, he had seen them frequently over the years in the immediate area. He would hear them moving around the river at night, but would not investigate for fear of his life. The animals, he confirmed, had a habit of surfacing under unwary canoes and spilling the occupants into the river. Like most Baka and other eyewitnesses, he was baffled as to why we were so keen to observe a *la'kila-bembe* for ourselves. "They are very dangerous; they will kill you!" he stated.

Brian felt that the immediate vicinity in and around the island was worthy of further investigation. We discussed the possibility of spending some time on the river after sundown. Unfortunately, night surveillance was out of the question because we did not possess night vision equipment and the nightshot feature on our Sony camcorders did not possess the range required to film any large dangerous creature from a relatively safe distance. Our plans were further frustrated by the fact that practically every villager living along this section of the river was simply too afraid to go out onto the river at night, especially when *la'kila-bembes* were still active there. We were disappointed, but still positive that we were in the right place at the right time for more research.

As the heat of the day gave way to the cool of early evening, Sando and Boima felt that it was safe to return to the river. We were still curious about the considerable upheaval of water we had heard earlier and used all our powers of persuasion to coax our guides to paddle around to the location of the disturbance. There, at the west side of the island, was a large breech, as if a large animal had forced its way inland, by perhaps thirty or so feet, to get at the fruits and leaves on the branches of the trees.

As we began the slow journey back to our camp, we also noticed more sizeable breeches along the riverbank on the east side, which could only have been made by an animal (or animals) as big elephants. This was surely an area where *la'kila-bembes* were active. Pierre had never seen these forced paths before, as he spent most of his time in the deep forests, but he was certain that his guides were correct. These were not made by any elephant that he was familiar with, especially as the river banks were very steep and offer little in the way of support for even the smallest elephant.

Later, we learned of another smaller island, much closer to the Congo side of the river. It too was known as a "forbidden area" where none of the local fishermen dared venture. Eager to explore this location, Pierre arranged for the guides to be ready to embark immediately after breakfast. We spent the rest of our evening catching up with our notes while Pierre whipped up steaming plates of spaghetti and spam with tomato sauce. Sitting by the campfire, we discussed the new information that was coming in from other eyewitnesses. It was considerably easier to collect information on the animals here than in the Congo. Although the Baka feared the *la-kila-mbembe* as much as their Congolese cousins did, they were still willing to offer us their knowledge of them, provided we did not try to place them in harm's way during our search. We also found that once we had assured our informants that our desire was merely to observe and film a *la'kila-bembe*, and not to harm or capture one of the animals, they were a little more at ease and willing to help us. Besides, if the animals were as rare as the pygmies claimed, any attempt to kill or capture a specimen could have catastrophic results for the entire species. Clear digital film of our quarry was all we desired.

Our plans to explore the "Forbidden Zone," as Pierre called it, were scotched by a thunderstorm accompanied by heavy rain that pounded the area all night, flooding our tent. Our clean, dry clothes were soaked through and the entire camp turned into a mud bath. Cold, soaking wet, and thoroughly ticked off at the weather, we donned our rainproof ponchos and began the laborious task of digging drainage channels around our tents. After mopping up the water in our tents, we hung our clothes along a makeshift washing line inside a mud-walled hut to dry them by a cooking fire while Pierre prepared a hot meal. The entire day was literally a washout, with no chance of getting on the river. The dark, brooding clouds eventually began to clear as the evening ushered in a brilliant starry night. I cranked up the FR300 emergency radio set to break the monotony of the evening. We managed to tune into Voice of America for the latest world news, followed by British soccer results! A one-minute crank of the emergency handle allowed us to listen to the radio for an hour. We retired to our beds at 9:00 p.m., hoping for a decent night's sleep.

The morning came bright and sunny, with no storm clouds to threaten us with more rain. The ground was already beginning to harden up, and a visit to the small Languoe River revealed that the water level had risen considerably. Our river guides had not shown up, so rather than wait for them we decided to embark on a brief exploration of the surrounding forest. Timbo Robert and Bruno Abolo led the way with their fearsome hunting spears, moving silently and nimbly along the forest floor. Brian and I tried to keep up, catching our clothes on the thorns and spikes that seemed to grow out of every tree and bush. As we entered a particularly dense and dark part of the forest, Pierre motioned us to stop. As we listened, we heard the sound of chimpanzees screeching and whooping as they leaped and tumbled around in a clearing just ahead of us. "Wild chimpanzees," Pierre whispered to us. "They can be very dangerous if provoked." I believed him. Gorillas were a different matter. Shy, reclusive, and usually passive, the western lowland gorilla would much rather avoid a confrontation, unlike the tribalistic, aggressive chimpanzees. Suddenly some colubus monkeys scattered in the trees above us as the chimpanzees got closer. It was time to leave—quickly. We had no firearms and even the Baka were getting nervous. Thankfully we were able to retreat and reach the forest track safely. Our clothes had become damp and muddy in the forest, prompting us to change before dinner. Brian had brought a solar shower bag, which heated the water via the sun when hung from a tree. This provided us with a welcome warm shower in our camp, which was much better than bathing in the river. Michel Ezeze, a trained pastor, stopped by to visit us with Norbert Nanga. Michel wanted to develop a ministry here, but the overwhelming dedication by the locals to witchcraft had intimidated him into silence. Pastor Nini had no such qualms and loudly rebuked the devil at every opportunity while preaching daily in the village to an ever-growing congregation. After dinner of instant mashed potato and canned ravioli, we drank mugs of Tetley's herbal tea before planning our next visit to the Dja.

News was circling the area that white men were at Langoue looking for the *la'kilabembe*, and more people stopped by to visit. Our river trip was again cancelled by a loud dispute between our two guides and the owner of the canoe, who had not paid them. The mother of one of our guides, who also happened to be the village drunk, embroiled herself in the melee, which only exacerbated the problem as more villagers joined in and began to shout and gesture at one another. Before the situation turned violent, Pastor Nini intervened, rebuking everyone sternly and loudly demanding silence while he gave up a prayer to the Lord to intervene and resolve the situation. It worked. The owner of the canoe abruptly ceased arguing and paid Sando and Boima for their services. The crowd dispersed, Pierre chose two new river guides, and Norbert Nanga kindly arranged for another canoe from his plantation for our use on the river. The day ended with a church service, this time in a wooden building that the village relegated for the purpose. The poisonous grip of traditional sorcery over the lives of the people was beginning to wane as Pastor Nini continued to preach fearlessly day by day, praying for people, laying on of hands, and even casting out demonic spirits. Even the atmosphere in the village began to improve significantly, as people seemed happier and more responsive towards us. Pastor Nini even asked Brian and me to address the new congregation that he had established. We both took turns to speak, telling the people what the Lord had done for us in our lives, which prompted Pastor Nini to conduct an altar call. Over 20 people came forward and received Christ into their lives. After ten days in Langoue, we had a church of over 30 people, with more attending the daily sermons of Pastor Nini daily. On Sunday, we baptized five new believers in the Langoue River, which was followed by a church service, which pastor Nini wanted to end with Holy Communion. Did we have anything that could be substituted for wine? I had brought along some powdered Gatorade to mix with our water, but it was orange flavored. "Can you check for me?" asked Pastor Nini. "Yes, I can, but it's still orange flavored," I replied. Entering the tent, I fished out the still unopened container of powdered Gatorade from the depths of my backpack and pried off the lid. I almost dropped the container when I saw that the powder wasn't orange. It was red. Somehow, my "orange" Gatorade was in fact a red-colored berry flavor. How was that possible? I didn't even want to think about it as I handed the container to Pastor Nini. He prepared the berry powder and some bread for communion, in which we all participated. One woman in the congregation kept making strange animal noises when Pastor was trying to preach, disrupting the service. Pastor Nini asked her to be silent, but to no avail. When she continued to make the odd vocalizations, the pastor firmly rebuked her "in Jesus' name" and she immediately lost her voice. For the next several days the woman simply could not utter a sound.

Although the rain continued on and off throughout the night, the sun broke through at dawn as we eagerly prepared to press on with our exploration of the mysterious islands that the natives feared so much. To our dismay, we found that the river level had increased by at least two or three feet, strengthening the currents and making our search a bit more hazardous. However, we pressed on and continued to look for signs of our water monster. Our new guides, Constantine and Celestine, both in their twenties, were from another small village nearby and were perfectly familiar with the area.

As Pierre directed our guides to swing around and head northwest for the "Forbidden Zone," they instantly became fearful and were most reluctant to approach the area. As we approached the island, I recognized it as the same location where John Kirk had observed a huge splash some two years before. Slowly and carefully, our guides steered the canoe into a narrow channel separating the island from the Congo side of the Dja. As we moved into the channel, Constantine and Celestine started to perspire, chattering nervously to one another. Before Pierre could ask them to stop talking, an enormous upheaval of water erupted almost directly behind our canoe. As we glanced back to the source of the disturbance, a large tree limb sprang upright, showering us with droplets of water after being bent almost double by some large, unseen creature as it hastily submerged. Constantine and Celestine were almost beside themselves with hysteria as we attempted to stop and examine the immediate area. In the wet season, the river encroaches inland by up to 100 feet, creating a sort of temporary swamp. This allows *la'kila-bembes* to move farther inland to feed while remaining partially submerged in the water. The dense bush provides very effective cover for them and thus makes it extremely difficult to spot any large animal that may be browsing under the cover of the overhanging branches and other vegetation. Unwilling to swing around, our guides instead paddled directly for the island, where we managed to grasp some overhanging tree limbs and remain stationary.

To our guides' dismay, they found themselves opposite a series of large caves located on a high muddy bank with the tops protruding just above the waterline. The caves were apparently used as lairs by the *la'kila-bembes* living in the area. Local fishermen avoided the area after finding their nets torn to shreds in the narrow channel. After spending an hour at the caves, hoping that our mystery animal might reappear, we abandoned our vigil, much to the relief of our guides, who made haste for the east bank of the Dja.

At the camp, a fisherman from the village of Mambele, by the name of Hinre Bossenga, visited us after learning about the purpose of our visit. In a very calm and matter-of-fact way, Hinre recalled an encounter he had in 1987, when a *la'kila-bembe* surfaced in the Boumba River directly under his canoe, breaking the vessel in half and sending Hinre into the river. As he looked back, Hinre could see his broken canoe sink into the depths of the river while the animal meandered about on the surface. He speculated that he may have disturbed the animal while he was casting his fishing net into a deep pool washed out at a bend on the river. He even mentioned that the animal possessed ridged dermal spikes about two inches or more in length running the length of its neck and back, which he believed were responsible for breaking his canoe in half. Hinre also knew of the narrow channel, or the "Forbidden Zone," and cautioned us about spending too much time there. More of the village people, especially those who were now attending Pastor Nini's church, starting coming forward to volunteer information.

Dolphine Monoumbou from Langoue once saw a *la'kila-bembe* in the Dja while accompanying her father on a fishing trip in 1987, and her husband, Paul, had also seen a very large specimen—he estimated its length at 70 feet!—in the Dja while fishing. More recently, another eyewitness, Marcel Bobas, observed a *mokele-mbembe* (as the animal is more commonly known near the border with the Congo where Lingala is spoken) in the Ubangui River, just north of Impfondo in 2000. All the eyewitnesses thus far who had seen the animals, sometimes on more than one occasion, spent most of their working day on the river. We asked at least three of the eyewitnesses to draw an image of the animals on the ground (without referring to the images in our binder). All three individuals drew pictures of what were clearly sauropod-like animals, possessing long thin necks, small reptilian heads, and long tails. Two of the witnesses included prominent dermal spikes with their images and one did not. The animals were described as being as big, if not bigger than elephants, reddish brown in color with caiman-like armored skin. Our Baka witnesses were Angou Jospeh, Bruno Abolo, Ande Yenda, Celestine Yenda and Timbo Robert. All of them confirmed other reports, particularly the intriguing zoological details of the animals. Pierre, who has complete faith in the trustworthiness of the Baka, stated that the pygmies fish in the rivers and swamps, and hunt in the depths of the vast forests, and are perfectly familiar with many strange animals. We cover these in other chapters.

The following day we headed straight for the island. We quietly paddled around the east side of the island and into the narrow, tranquil channel of the Forbidden Zone. As we passed the location where a huge unseen animal abruptly submerged behind our canoe three days before, we could see the breech that it had made as it pushed its way through the thick, tangles mass of foliage to reach its food supply.

Constantine and Celestine grew nervous again as we entered the channel. The caves came into view on the Congo side of the river, and we silently guided the canoe under cover of the overhanging branches of the trees on the island, which has turned into a muddy swamp during the rainy season. We named this "Swamp Island," which, along with the "Forbidden Zone," sounded like something out a low-budget horror movie. We almost expected the Creature from the Black Lagoon to surface and drag one of us down into the murky depths. The intermittent rain showers over the past two days had further increased the depth of the river and covered the tops of the caves. At 12:30 p.m., a sudden rainstorm hit the area and pounded us mercilessly. Our guides struggled to keep the canoe on course as we took on water during the slow, torturous paddle back to our landing point. Dejected and soaked through, we trudged back to the village and slumped into our camping chairs, which had been moved into a mud hut, where Pierre prepared some hot soup to lift our spirits. The rest of the day was literally a complete washout. Perhaps tomorrow would be better.

The morning of Thursday, October 30, was to be our last day on the river. Pastor Nini boldly proclaimed that this would be the day when we would encounter the *la'kila-bembe*. The Lord assured him that this was truly the day, he stated in his matter-of-fact way. However, our guides refused to venture back to the Forbidden Zone, and only wanted to stick close to the bank where we could land quickly if we were fortunate enough to encounter a *la'kila-bembe*. Pierre tried to reassure them that we were also interested in looking for other animals, and we would pay them more money just to take us on the river for the day, with perhaps a brief visit to the Forbidden Zone. Reluctantly, Constantine and Celestine climbed into the canoe and started paddling south. We stopped at a Baka camp in order for Pierre to talk to the people about our river monster. As was the usual custom, we were welcomed with the quiet graciousness that is literally the trademark of the Baka. After being seated on two small wooden stools, our hoist showed us some of the small game that they had hunted that morning. One curiosity took us by surprise. It was a small tusk that apparently belonged to a sort of wild pig that, according to our informant, lived underwater. At first I thought that our hosts may have been talking about the pygmy hippopotamus (*Hexaprotodon liberiensis*), but these are only found in Liberia, Sierra Leone, and the Ivory Coast. However their possible range may extend into Nigeria and Equatorial Guinea, which could theoretically allow them access to the river system of Cameroon. We left the mystery behind for another time as we headed back to the river and on to another Baka village. This time we remained in the canoe

Congo Bill in His Element

Exploring Lake Tele

while Pierre walked to the village. They did not often see white people and Pierre thought it best to approach them first to seek information on the *la'kila-bembe* alone.

An hour later, Pierre returned with some interesting information that certainly gave us food for thought. According to his informants, the animals spent most of their time browsing along the river's edge hidden by the overhanging tree branches and tangled foliage. The animals emerged from the deep pools in the river around the hottest part of the day, but could be observed occasionally as late as 3:30 p.m. in the afternoon, before the day begins to cool. Another interesting piece of information concerned a loud crack or slapping noise that the animals made with their tails by striking the water. This frightened the Baka considerably and kept them away from the river until they were sure that the animals weren't around. This new information intrigued us considerably. Could the tail slapping episodes be some form of territorial declaration, or perhaps a way to attract a mate? By 1:30 p.m. we headed for the Forbidden Zone, much to the dismay of Constantine and Celestine. As before, we slowly approached the island from the east before stealthily entering the channel from the south. Rather than guide the canoe back into the foliage on the east side of Swamp Island, Pierre instructed our guides to stay in the center of the channel. The three caves on the Congo bank were still almost completely submerged, which made it impossible for us to determine if they were inhabited by any large creature. This was our last shot at discovering anything significant on this expedition. Pastor Nini had confidently predicted that this was the day that we would encounter our fabled river monster. So, where was it? As Brian and I focused our attention on the left and right banks of the channel, Celestine suddenly gave a cry of alarm and started pointing directly ahead in the channel while speaking rapidly to Constantine, who was at the back of the canoe. Pierre joined in on the rapid-fire conversation and turned towards us. "There is a very big animal crossing the river just ahead of us."

Brian and I panned our camcorders left and right, using our zoom lenses to try and zero in on the mystery animal. Pierre stood upright, stared directly ahead of him, pointed to an object in the river, and gave a cry of astonishment. Constantine and Celestine were now in a panic, trying to turn the canoe around. Pierre loudly ordered them to continue north through the channel. Our guides were almost hysterical with fear, but Pierre urged them on while remaining upright in the canoe, which threatened to capsize at any moment. We were almost clear of the channel, passing the submerged tree log with only its tip protruding out of the river. Once we were out of the channel and beyond the northern tip of Swamp Island, Pierre sat down and promised to explain more once we were back safely on dry ground. Brian and I joked earlier on the expedition that it would be interesting to see just how fast two pygmies could paddle a canoe if they encountered a *la'kila-bembe*. If Constantine and Celestine paddled any faster, they would have set an Olympic record.

Back at the village, we sat down and discussed the episode with Pierre. During our excursion through the Forbidden Zone, Celestine spotted the large, reptilian head of a *la'kila-bembe* poking out of the river as it slowly crossed the water from one of the submerged caves to Swamp Island. "How did he know it was a *la'kila-bembe*?" asked Brian. "He could see that big body just under the surface," replied Pierre. Constantine was also able to observe the animal, but at 6 feet 2 inches tall, Pierre got the best view of the head of the monster. Pierre described the head as being very much like a large python, but twice the size, and darker in color, almost reddish brown. The eyes appeared to be typically reptilian, and he was certain that the animal was walking along the bottom, not swimming. We estimated that the river was around twenty feet deep at the time, which would have given our mystery animal a head height of at least the same. Pierre, Constantine, and Celestine observed the head of the animal from a distance of about 50 feet. As the animal approached Swamp Island, it became aware of the noise made by our frightened river guides and dipped its head underwater.

They were absolutely certain that the animal was not a snake, a turtle, a crocodile, or any other kind of swimming animal, as the strong current would have affected it passage as it moved across the river. The size of the animal's bulk (as observed briefly breaking the surface), and its movement across the river convinced Pierre and his co-witnesses that the animal was walking along the bottom. All three men are highly experienced hunters and perfectly familiar with all the animals of the Dja River region, and Celestine and Constantine spend a good deal of their time on the river, where they regularly observe snakes, crocodiles, turtles, and other large animals. Elephants are only present on the upper reaches of the Dja River, especially around Nki National Park, about 75 kilometers north of Langoue, and in certain parts of the Boumba. Hippos are never found around the confluence of the Dja, Ngoko, and Boumba Rivers, and are only found in the upper Dja around the Dja Faunal Reserve area. We found that hippos and elephants are completely absent from those areas of river and swamp where *la'kila-bembes* are reputed to be active, and the local fishermen never saw them anywhere in the lower Dja or Boumba Rivers. We were elated. So close, yet so far from bagging the evidence—at least on camera. But we now know where, when, and how to look for these extraordinary animals.

On Saturday, November 1, we packed up our camp at dawn and waited for the Toyota pickup truck to carry us to the Boumba and then onto Moloundou. It was hard to say goodbye to the friends we had made at Langoue, but we hoped that it wouldn't be too long before we were back. Our transportation did not arrive at 6:30 a.m. as pre-arranged, and Pierre was anxious to get underway. We were still waiting by 7:30 a.m., so Pierre decided to walk the 8 kilometers to the Boumba River, where the cable/pulley mechanism that operates the ferry had been repaired by two French loggers.

Eventually a wheezing rusted-out Toyota pickup truck arrived to collect us. After piling our equipment in the bag, we slowly chugged back to the Boumba, where Pierre was already on the way back with our safari bus. By 10:00 a.m., we had breakfast in Moloundou and hit the road for Yokadouma. The appalling muddy track bogged our vehicle down twice as we slowly navigated our way around deep pools and at least six logging trucks that had overturned on their sides, spilling their haul of hardwood timber onto the road. We finally reached Yokadouma and the Elephant Hotel after dark. The electrical supply was still disconnected, and we signed in at the front desk by candle light. The manager then announced that the "rules" had been changed and that two men were no longer allowed to share one hotel room, otherwise we would have to pay double the price. We knew full well that this was just another scam to extract more money from white guests, so we simply grabbed our bags and headed back to our vehicle. The manager was unwilling to change his mind on the price and did not even attempt to persuade us to stay. Pastor Nini instead suggested that we book in at the Auberge Hotel located in the centre of town. The difference was immediately apparent. The power supply was connected, and the rooms were clean and even possessed ceiling fans and television sets! The hotel restaurant served chicken with fried plantains and sliced bread with mayonnaise. Although the chicken tasted a little off, we tucked in while watching an African soccer match on TV. Although everyone got a decent night's sleep, we all paid the price the following morning for eating chicken that was clearly unfit for human consumption. Everyone dashed for the bathrooms as our innards felt like they had turned to liquid. But, we had a long journey ahead of us and we needed our strength for the journey.

We forced ourselves to eat a large breakfast of omelettes with bread and coffee before buying some supplies for the road. We left Yokadouma at 9:30 a.m. and pressed on for Bertoua, stopping at Batouri to allow our Muslim driver to pray. I took the opportunity to use the latrine at the safari bus stop and was charged twice the fee of 25 cents by the Muslim attendant. "Infidels" were obliged to pay double!

By 5:30 p.m. we finally made it to the Anderton home in Bertoua. We dumped most of our camping gear in the storage container while Pierre visited La Mission Catholique to book our rooms. The nuns at the mission were, as usual, very accommodating and we were billeted in a clean, spacious room each. After a refreshing shower and a change of clothes, Brian and I walked to the Restaurant Baron, which is owned by an elderly Spaniard who has lived in Cameroon for over 30 years. We enjoyed a delicious steak dinner with vegetables, and a glass of wine, followed by a solid night's sleep. Our remaining three days were spent at the Anderton's place cleaning our tents, washing our laundry, and sending emails to our family, including an update to Milt Marcy.

Our final drive back to Yaounde went without incident, and we departed on our Swiss Air flight for Zurich, then Chicago, and finally to Calgary. Brian later caught the small commuter flight to rejoin his family in Regina, Saskatchewan. After a full 24-hour rest to recover from the long journey, I called Milt Marcy and discussed the expedition in detail. We were able to confirm that the *la'kila-bembes* were well known in the area, and that they were living animals that had been observed recently, quite apart from our own three "close encounters." Milt agreed that it was imperative that we return to the Forbidden Zone as soon as possible. How soon another expedition could be mounted was another matter.

> Be sure, if you go against *mokele-mbembe*, you will die!
>
> —Baka elders,
> Langoue village

8

RETURN TO THE FORBIDDEN ZONE

Although a second exploration of the Forbidden Zone and its immediate surroundings was imperative, I went over my notes carefully to try and make sense of the relationship between *mokele-mbembe* and the seasonal fluctuations in southern Cameroon. The animals are most active in the wet season between September and January, but are hardly ever seen in the dry season. Several *la'kila-bembe* and *mokele-mbembe* eyewitnesses stated that the animals are sometimes observed digging out their caves and clearing out debris washed inside by the river. If we wanted to observe one, we would have to return in January or early February at the latest. This confirmed earlier information that the veteran Congo missionary, Gene Thomas, had gleaned from his conversations with the Aka pygmies over the years. The animals were most active in the wet season, but hardly ever observed in the dry season, especially after the month of February.

Milt Marcy was still able to offer some financial support for another trip to Cameroon, and Brian was willing to return in early in 2004. I was unable to travel again so early due to work commitments, but Milt had someone in mind who was keen to join our team. Peter Beach, a microbiologist and an old friend of Milt's, had taken an interest in our work and was willing to join Brian for a second trip to the Forbidden Zone. The following account is from Peter's notes, edited here with his kind permission.

On February 6, 2004, at 8:25 a.m., Peter and Brian left Portland, Oregon, bound for Africa. The two biggest concerns for them were the lack of government documentation, and their limited funds. However, Paul Nkeng remained keen to assist us in any way possible and requested that Brian and Pete provide a film projector for movie nights at the main church in Yaounde. Pete's journey was fraught with difficulties from the beginning, which only grew more complicated along the way. Pete was delayed at the security check due to a small pair of scissors in his backpack. Brian passed through with no problems, and decided to meet Pete at the boarding gate. Unfortunately, the delay cost Pete his flight from Portland to Dulles International in Texas. He and Brian were already hundreds of miles apart on the first leg of their journey. Although he managed to secure a later flight to Chicago, Pete found that he did not possess a paper ticket to Paris, but his luggage and ticket to Yaounde were in order. A last minute spelling error (Beech instead of "Beach") on Pete's ticket had been cancelled out, but another had been thankfully provided and Pete was able to restart his heart again. The flight from Chicago to Paris was delayed due to a technical fault, delaying the departure to Paris by one and a half hours. Finally, the United Airlines flight to Paris got underway and was assisted by a 50-70 mph tailwind, giving the airliner a cruising speed of 600 mph, which helped to make up an hour and a quarter. There was a chance that the two explorers would catch up in Europe, as Brian was by this time in Paris awaiting the connecting Air France flight to Yaounde. After landing in Paris, Air France would not let Pete board the same flight that Brian was already seated on at the departure gate because he didn't check in one hour before. Pete dashed off to the Air France African departures check-in, which was located in a dingy little corner of Charles de Gaulle International Airport. The airport was four times the size of the one back in Portland, and as everybody spoke French, Pete's language barrier only delayed him further. In Pete's own words, "Everybody looked American but spoke only French."

After Pete explained his predicament to the French check-in clerk, the man called his supervisor, who turned out to be a very pleasant woman but whose command of the English language was as limited as Pete's French. After a fusillade of verbs and hand waving, the woman finally grasped the situation and took off with Pete in tow. After negotiating what seemed to be a rat's maze designed to break the spirit of weary airline travelers, they stopped at terminals 1A, 2A, 2B, and then 1B. Fifteen minutes later they ended up back at the Air France desk. After re-examining Pete's ticket, the lady thought that she knew the source of the problem and beckoned Pete to follow her. So off they went again on another jolly jaunt through the worst airport ever designed and ended up back at the United Airlines desk. The clerk there said that nothing was wrong with Pete's ticket and he should return to the Air France desk. As it was United Airlines' fault that Pete was late arriving in Paris, they were responsible in ensuring that he completed his last leg to Yaounde in a timely manner. It was clear that Pete would not able to board the same Air France flight to Yaounde that Brian was on, so he would have to find another flight. By this time, Air France was willing to accommodate Pete, but the next Air France flight to Yaounde was seven days away. To make matters even worse than they already were, Pete found that he couldn't get an exchange ticket with Cameroon Airlines, because his discount airline ticket did not qualify for an exchange or a refund. In spite of

United Airlines' promise to get him to Yaounde, they left him high and dry. The only two options were to wait in Paris for seven days and fly to Yaounde at no additional cost, or purchase a ticket with Cameroon Airlines and leave the same day. Milt had dug even deeper into his own pocket to finance the expedition, and Pete did not want to delay his progress anymore than absolutely necessary. For some odd reason, a one-way ticket to Yaounde from Paris cost $2,369.00, but they could offer him a return flight for $1,016.00! By this time Pete was exhausted and didn't want to know why. But was Air Cameroon willing to redeem the still-valid Air France ticket, while Pete would pay the difference? "Non!" stated a very officious French female employee, who pointed out a paragraph in the rules and regulations of the airline industry to her superior. In the end Pete had little choice but to bite the bullet and pay for a second ticket with Cameroon Airlines.

Reluctantly, he forked out the extra money for a return ticket on a reasonably airworthy Boeing 747-300, and wondered if he had just been conned. As a final insult to Pete's extraordinary patience, his flight took off 40 minutes late without a word of explanation. By 6:45 p.m., Pete was flying over Africa at 30,000 feet just as the sun was dipping beneath the curvature of the earth. Although he hadn't slept for almost 30 hours, this humble scientist and child of God felt energized as he marveled at the beauty of the planet: "a layer of deep salmon, gray cloud, a cream-colored layer, then light blue blending into a deeper hue."

At 10:45 p.m., the Cameroon Airlines 747-300 touched down in one piece, delivering Pete safely on African soil. Brian had arrived three hours ahead of Pete and had gone straight to his hotel with Pierre Sima. Unfortunately, Pierre wasn't there to meet Pete and he didn't know the name of the hotel where Brian was staying. The message that United Airlines promised to convey to Brian about Pete's alternative flight arrangements had apparently not reached him. Pete left the airport with Kamani Bernard, a pleasant and trustworthy local cab driver, and proceeded to hunt for Brian. After checking sixteen hotels without success, Pete checked into a modest establishment for the night. He needed to rest and regain his strength before commencing with his search.

By 6:30 a.m., after barely four hours sleep, Pete managed to call Milt in the USA from the hotel front desk. Milt was relieved that Pete had made it to Cameroon safely, and provided him with the name of Brian's hotel and his room number. One hour later, Brian opened his hotel room door, astonished to see that Pete had made it to Africa so quickly. It was Sunday morning in Cameroon, and practically everything in the city was closed. As they could not change any money, Brian, Pete, and Pierre decided to relax and go over their plans for the expedition. Pastor Paul Nkeng sent a driver to pick up the team for the morning service. The driver happened to be a serving colonel in the Cameroon Air Force. He was trained to fly the McDonnell Douglas F-15 Eagle fighter, but flew the far less exciting Dassualt-Breguet Alpha Jet, a small two-seat trainer and ground attack aircraft.

Pastor Nkeng welcomed the team warmly and was delighted to receive the promised film projector for his church. The projector, however, needed speakers and so these were added to the shopping list of expedition supplies, along with new underwear for Pete as his luggage had still not arrived from Paris.

After a lively church service, Pastor Nkeng invited the team to his neat and tidy single story home for lunch and a televised soccer match between Cameroon and Nigeria. It was always wise to pray for the host nation's team to win, lest riots in the streets ruined your stay there! Although Pierre had planned the trip well ahead of time, for some odd reason Paul Nkeng wanted Brian and Pete to travel to Bertoua by train, or alternatively, by scheduled bus, without explaining why. Either way would have taken twice the time and cost twice the money. Pierre's arrangements won the day and the last day in Yaounde would be spent exchanging American dollars for Cameroonian francs.

By Monday afternoon, all the supplies had been purchased, including two new speakers for the church projector. Brian had also acquired a bicycle for Norbert Nanga, who had requested our assistance in securing a bicycle to assist him in his travels between the villages that lay along the narrow paths that ran along the Dja River. Pete was amazed that four adults, three boxes of groceries, two large speakers, and a bicycle could all fit into a single Toyota Corolla taxi cab.

Although the team was ready to depart for Bertoua the following morning, Pete's luggage had still not arrived from Paris. United Airlines had originally routed Pete's luggage through Air France from the United States, then Air France in turn re-routed the luggage through Cameroon Airlines in Paris at the last minute, but did not follow through to ensure that everything would arrive in Yaounde on time.

Finally, Pete's luggage arrived at the airport. By 9:30 p.m., everything he needed was ready for the journey to Bertoua the following morning. As usual, things do not always go according to plan in Africa, and the team eventually left Yaounde at noon, arriving in Bertoua by 5:00 p.m. The police stopped the safari bus on a routine check and found shotgun shells in the vehicle. After arresting the driver, the rest of the team was marched down a narrow path in the bush. Thankfully, the path lead to a police station rather than a firing squad, where the driver was given a fine of 10,000cfa, which Pierre added to the cost of the trip, totaling 100,000cfa, as a gesture of good will, lest he lose influence in the town. Finally, the expedition equipment and camping gear were collected from Pierre's plantation where everything had been moved. Joe and Lillian Costillo were the new missionaries in Bertoua, replacing the Douste family, who had completed their tenure in the field and had since returned to Canada. Brian and Pete booked rooms at the *La Mission Catholique* in town where our dear friend, Sister Marcellin, was stationed.

Wednesday, February 11th was Youth Day in Cameroon, and most of the stores in Bertoua were closed for the day, which was marked by parades and speeches. This delayed the team's departure by yet another day, as they would not be able to purchase additional supplies for the trip, including roofing materials for the Baka church in Langoue. It was wise to get used to delays in Africa as they seemed to pop up almost daily and in all circumstances.

Finally, on Thursday, February 12, the team left Bertoua for Moloundou with Pastor Joseph Nini as the fourth member. The rainy season was over and the roads were no longer swimming in mud, but had hardened into the corrugated surface that shook and rattled every vehicle on the road. After an overnight stop in Yokadouma at the Catholic Mission, the remainder of the journey south was frustrated by a collapsed bridge which added another two hours to the journey via narrow jungle track.

One final night was spent in the notorious "Moloundou Hilton" before finally crossing the Boumba River early on Sunday morning. As the team prepared to move down to the river, Pastor Nini mentioned to Brian and Pete that he had a strange dream the night before. A large solidly build man that he somehow knew was in charge of the expedition, appeared to him in a dream, and told him to instruct the two explorers to search the river for *mokele-mbembe*. He also asked the pastor to pray by the river when Brian and Pete were exploring the Dja. I have learned not to discount such stories, as Brian had mentioned to me before that members of the prayer team in his church in Regina were awoken in the middle of the night with a sense of urgency to pray for us. Later, we found that the times of these occurrences happened during our river exploration of the Dja. Pastor Nini had never met Milt Marcy or knew what he looked like, yet accurately described him from a dream. Before the final leg of crossing the Boumba, the team visited the chief of police, the district commissioner, and the mayor. As it was customary to present gifts to certain officials, each one was given a hunting knife, a pair of binoculars, and an electric shaver.

After crossing the Boumba on the floating platform, the team finally made it to Langoue and were greeted by almost the entire village. Pastor Nini called a church service together, which was followed by eight baptisms in the Langoue River. The new believers included Chief Nicole Daniel, Nobert Nanga, and Orel Nanga. The latter is now serving as the elder of the new village church. Norbert was delighted to receive his new bicycle, and pledged the unlimited use of a large canoe in his possession as often as we needed it for river exploration, and without charge. During a church service later in the day, a thunderstorm swept in, pounding the metal roof of the little schoolhouse that doubled as the village church, deafening the 25 attendees, but the service went on regardless of the weather.

At 10:00 a.m. on Monday morning, Brian, Peter, and Pierre made their way to the Dja River. At the landing point, Norbert Nanga had beached a large canoe with the water bailed out and ready to go. There were no fearful river guides on this occasion, as the team knew exactly where they needed to go: Swamp Island. Pete volunteered to sit at the rear of the canoe and paddled towards the island.

Pierre sat in the middle and Brian was positioned at the bow, looking for a suitable place to land. It was immediately apparent that the river level had decreased considerably. Leaving the canoe on the sandy bank, Brian and Pete immediately began filming their progress as they walked quietly across the sand bar and splashed through the shallows. The tangled mass of bushes, branches, vines, and thorns made it difficult to reach the western side of the island without making any noise. As they came closer to the bank that bordered the Forbidden Zone, the team expected to see a *la'kila-bembe* at any moment. Pierre pointed across to the caves on the Congo side as they moved past them, one by one. Each cave was sealed from the inside. Were the animals present in the enclosed chambers?

Pete carefully measured the caves and found them to be 15 feet by 15 feet at water level, with five feet of water remaining in the channel, separating the island from the Congo side of the river. Upon closer examination, Brian and Pete were able to determine that the caves had been sealed from within, with small air vents near the top. As they continued their examination of the area surrounding the caves, Pete found some odd grooves and marks on the bank. The once soft mud had hardened, preserving the marks perfectly. The marks weren't just odd—they resembled claw marks where an animal of considerable size had scooped out mud and other debris from the immediate entrance to the caves and the area surrounding them. Pete had left some plaster of Paris at the camp and was eager to start making some casts. After checking with the village, it appeared that no one had been near the Forbidden Zone since November 3rd 2003, when Pierre and our river guides briefly observed one crossing the river. Pierre did visit the area on January 7th, 2004, and found one of the three caves had been sealed by that time. He wanted to examine the remaining three caves, but the Baka hunters who were with him at the time were deathly afraid to remain in the area, and Pierre had to abandon his plan to camp out on the island. Further information was forthcoming about the tail-slapping episodes that could be heard from a considerable distance on the river. The extreme activities of the beasts, including mating and tail slapping, apparently occur September to October, supporting similar information gathered from other eyewitnesses and informants from different locations on four other river systems in Cameroon and the Congo.

By daybreak on Tuesday morning, Brian, Pete, and Pierre were ready to leave for the Forbidden Zone, but word reached them that the *Chef de Politique*, Joel Nanga, was scheduled to arrive at their camp at 9:00 a.m. They had missed him in town but left gifts for him before pressing on to Langoue. The Commissioner eventually arrived at 12:00 noon, which was a common ploy by African officials to ensure that everyone would be awaiting his arrival. He took up the best part of the afternoon, as most African bureaucrats tend to do, scotching the team's plans to return to the target area early for a full day's research. Yet another frustrating delay had stopped the team's vital research from making further progress. The claw marks that were left on the high bank, coupled with the tranquility of the island, could reveal further evidence. Hopefully, the next day would be considerably more productive and advance the team's progress towards more substantial results.

On Wednesday morning at 6:46 a.m., the team quietly paddled their way across the river to Swamp Island, one half mile southeast of their embarkation point on the Dja. Shortly after landing on the island, Pete made sure that he recorded the precise location of the caves on his GPS system. On this occasion, the team made a measuring stick to accurately gauge the height and width of the caves. Pete got to work with the plaster, working carefully and diligently to make casts of the claw marks that were preserved around the caves. Later, the team photographed and measured the caves, which they confirmed as 15 feet in height and at least 15 feet in length, and dug into a bank measuring 20 feet (taking into account the remaining five feet of water in the channel). This discovery reveals remarkable similarities with an open cave discovered on the Likouala-auxe-Herbes River, south of the town of Epena, in the northern region of the Republic of the Congo, and shown to Roy Mackal and his team in 1980. The local tribes people claimed that the cave was used as a lair by a *mokele-mbembe* in 1979.

During the first-hand examination of the west bank of Swamp Island, a large breech was discovered almost immediately opposite the location where a suspected *la'kila-bembe* was observed crossing the river from cave #4 to the island. Further investigation found that leaves were stripped from the branches to a height of 18 feet, with a series of large, elephant-like footprints impacted in the mud directly beneath the feeding area. Pierre was delighted with the find and was emphatic that upon close examination, the elephant-like footprints were made by an animal the same size, if not bigger than an elephant, but were markedly different. Having tracked literally hundreds of elephants through forests and across savannahs, Pierre was simply unable to reconcile them with any known large creature that he was familiar with, including the hippo. With great care, the team measured out the prints, which were placed in an elliptical pattern, revealing that the animal had been moving left to right and vice versa as it browsed on the leaves of the trees on Swamp Island, but remained close to the water. The prints, which measured 12 inches across and 14 inches in length, were impressed into the mud by one quarter of an inch. At least seven of the prints were relatively clear, with others were obscured as the animal had walked on an area that was covered with leaves. At least two of the prints were impressed into the mud slightly deeper than the others, suggesting that the animal sat on its haunches briefly, or shifted the bulk of its weight to its hind legs while attempting to reach the higher branches. To Pierre's trained eye, this was a very unusual animal indeed, and Pete used his biology training to use the size, disposition, and dispersion of the footprints to estimate the size of the animal, and later make comparisons to other large animals, as we shall see.

The end of the wet season was around mid-November, with the occasional downpour. The river levels decline rapidly after this at a rate of about one foot a day, not including the occasional rain shower. Factoring in a few downpours in the weeks following the end of the rainy season, it takes between two and three weeks for the water level to drop to its level of 4 to 5 feet before the end of January, revealing the cave mouths. This means that in the month of December, when the food ran out and the water level was dropping fast, the animals went into their caves and closed themselves in rather rapidly.

Later at the caves, Pete left a small tape recorder lodged into the air vent of cave #3 in an effort to record any kind of movement or vocalizations within. As he was engaged in making a plaster cast of one of the claw marks, Pete was startled by the distinctive sound of something scraping against the wall in one of the caves. The sound was so clear that he even asked Brian and Pierre if they had heard the same noise, even though they were in conversation while examining another cave. After Pete drew their attention to the noise, Brian and Pierre could also hear a distinctive scraping sound, as though something was attempting to claw its way out of the sealed chamber. Pierre at once became concerned for the safety of his team members and ordered everyone into the canoe. He reasoned that if a charging elephant is the most dangerous animal he has ever known, then he did not want to jeopardize his life or the life of his colleagues by facing an ill tempered *la'kila-bembe* that was able to kill any elephant or hippo with ease. Nobody wanted dead white men on their hands in Africa!

Later at the camp, Brian and Pete listened carefully to the tape recorder too see how well the scraping sound had been picked up. The microphone has originally been left about two feet inside the air vent (with the microphone facing inwards) at cave #3, while Brian and Pete continued with their examination of cave #4. Surprisingly, the microphone was acute enough to pick up Brian and Pete's voices from cave #4. This suggested that the caves were in fact tunnel entrances, with the distance from cave #3 to cave #4 measured at 50 feet, and from cave #1 to cave #2 the same distance apart. The distance from cave #2 to cave #3 was 300 feet. The caves were interconnecting chambers that, given their size and distance from one another, would have been sufficient to house at least three large *la'kila-bembes* with their offspring. A final examination with measurements and photographs taken of the footprints on Swamp Island completed the expedition. As Pete and Brian only had church papers to guarantee their freedom of movement in the country, they could not afford to arouse suspicion regarding their river excursions, and decided to play it safe by completing their investigations on Friday, February 20th. It was a real breakthrough in our search for these animals.

The journey back to Bertoua commenced early on Saturday morning, requiring only one overnight stop at Yokadouma. The team arrived in Bertoua on Sunday with rooms at *La Mission Catholique*, waiting for Brian and Pete. Time was rapidly running out for the team, and Pierre quickly arranged for further transportation to Yaounde, before arranging dinner at his neat and spacious home for the team's farewell.

The last leg of the journey to Yaounde was completed without incident on Monday, February 23rd, where a final stay at *Le Mercure* was arranged prior to the flight home, which was scheduled to depart on the following Friday. Pete's hand-luggage bag was again searched at the Yaounde Nsimalen International Airport. The first security check turned up the plaster casts.

The security officer kicked up a fuss and would not let Pete take them out of the country. Pete protested and stated that they were only souvenirs. The man suggested a price in a low voice to avoid attention. Pete didn't quite have the amount of currency in a single bill that the official suggested, but parted with a larger amount. Pete was waved through but almost immediately encountered another security officer, this time a woman who also searched his bag, She ignored the plaster casts he had made of the alleged *la'kila-bembe* claws but took a curious interest in his Bible. After another heart-stopping moment, he was waved through and boarded the Air France flight for the long flight home via Europe.

The expedition was now officially over, but progress had moved our search forward in leaps and bounds. Pierre was greatly encouraged by the evidence that the team had gathered at the Forbidden Zone, and decided that the area was worth visiting again—funding permitting. To that end, he committed himself to visiting the location sometime before or during the rainy season to see what possible progress he could make in the search for the much feared but elusive *la'kila-bembe*. The big question was, would he be able to crack open the mystery wide enough to make a difference?

Pierre and the Forbidden Zone

Pierre's curiosity about the sealed caves at the Forbidden Zone got the better of him, and he decided to take another look at them in June-July 2004. Pierre traveled to Langoue in a hired vehicle and left it at the village with a guard. During the journey they stopped in Welele to pick up nine Baka trackers before pressing on to the village of Langoue. Unfortunately, Norbert Nanga was not at his plantation, so Pierre arranged to hire two other canoes from another fisherman. The first canoe was to convey Pierre and his team across to the Congo side of the river, and the second canoe remained with them in the event of an emergency. Pierre divided his group into two teams of five people. Pierre and his group remained camped close to the Forbidden Zone, while the second team camped near Moloundou, not far from the confluence of the Boumba and Ngoko Rivers. They maintained nightly vigilance for five days in bright moonlight, sleeping only during daylight hours. On the sixth evening, the moon vanished behind a thick layer of cloud, thus robbing both teams of valuable nighttime illumination. At 11:00 p.m. on the sixth night, a canoe with a bow lantern was spotted slowly entering the Forbidden Zone from the north. Pierre challenged at once, enquiring why they were putting down their fishing nets in the channel. Were they not afraid of the animals that lived in the caves? The fishermen replied that noisy, motorized canoes had been moving through the area in March and April, and had not been menaced by the animals, therefore they concluded that it was safe again to fish in the channel. Pierre did not want the tranquility of the area to be disturbed and tried to discourage the fishermen from entering the channel. However, the fishermen laid out their nets and caught plenty of fish before moving to another location. During daylight hours Pierre found that the caves were still sealed, therefore, the animals could not have moved away, but were merely holed up in their caves, awaiting the advent of the rainy season. This area of southern Cameroon experiences two rainy seasons each year. The long dry season is from December to March, the short rainy season from March to June, the short dry season in August, and the long rainy season from September to December. The animals have been seen only very rarely during the short dry season, but are most active between September and mid-January.

During his surveillance of the Forbidden Zone, Pierre was visited by an elderly Baka man named Djouma, accompanied by his wife, Loma. Djouma was well known in the Baka community for his wide knowledge of the flora and fauna of southern Cameroon and Northern Congo, including the *la'kila-bembe*, but did not think that they were any more unusual from the other animals that he was familiar with, except they were more dangerous if disturbed. He explained that when a female *la'kila-bembe* gives birth, it changes its location to find a shallow swamp where it will clear out a small area to rear its young. Djouma further explained that if the animal leaves an area and does not return, it leaves the caves open. This way, the water will fill the cavern and collapse the structure from the inside. If the animal intends on returning to its original location, it will seal the cave entrance and continue to browse along the river and/or reproduce. The cave-hollowing activity of the animals occur between November and December, which corresponded almost exactly with the findings we had made during our first initial investigation of the Forbidden Zone in November 2003. With this new information, Pierre decided to move from his location and conduct further research downriver in the Congo and nearby swamps. On the very morning that he was preparing to leave the Forbidden Zone, a huge splash emanated directly from one of the caves in the Forbidden Zone. As the Baka were too afraid to investigate the disturbance, Pierre went alone to the spot and found that one of the caves was empty and had collapsed. Where did the animal go? Could it have been sufficiently disturbed by the fishermen in the channel to vacate the area? As the water level was still fairly high, the animal could have gone anywhere, with plenty of places to hide in the rising river and remote swamps. As the moonless nights made it more difficult to conduct surveillance without sophisticated night-vision equipment, it was time to try a different approach.

After collecting his truck form Langoue, Pierre crossed the Ngoko River via vehicle ferry and was promptly forbidden to enter the Congo with his Cameroon registered vehicle. After much haggling and agreeing on a "fee," the Congolese border guards allowed Pierre to cross the border and proceed to Sembe, a pygmy settlement located on the Sangha River. Pierre found that the pygmies were very knowledgeable about the *mokele-mbembe* and its general habits, but no new information was forthcoming other than the fact that they see them in the Sangha north of Ouesso from time to time. From Sembe, Pierre enquired at Souanke before eventually returning to Ouesso. After completing his enquiries in the Congo, Pierre came to the realization that while the animals are also well known in the Congo, particularly along the Sangha River, many of the people were more knowledgeable about the animal than they were prepared to reveal, and outsiders enquiring about the animals, whether western or Cameroonian, were simply not welcome. He concluded that the search for *mokele-mbembe* would yield far more promising results in the river system of southern Cameroon where the animals had been seen frequently and recently, and where the people were far more helpful in providing specific information on dates, times, and locations relating to the movements of the animals.

> Hippos are curiously absent from areas where *mokele-mbembes* are present.
>
> —Roy P. Mackal

9

Milt Marcy Ventures Forth

Having funded the previous two expeditions to Cameroon, Milt Marcy felt encouraged enough by Brian and Pete's findings to directly participate in the search for *mokele-mbembe*. On January 11th, 2005, Rob Mullin flew out of Kansas City International Airport on schedule, and landed in Newark, New Jersey, only to learn that Milt Marcy and Peter Beach had been delayed in their departure from Portland, Oregon. Rob would have to continue to Africa alone and meet his fellow explorers there in a few days.

After landing in Yaoundé on the morning of the 13th, Rob was met by Pierre Sima and was billeted at a missionary house in the city by Minnie Stoumbaugh of CABTAL (Cameroon Association for Bible Translation and Literacy). Milt and Pete arrived late on the 14th, and true to Air France's reputation of losing baggage, did not deliver Pete's luggage until the 16th. A meeting with the current and very pleasant Minister of Science, Madeleine Tchuenté, ensured that the team received their official documents in a timely manner.

Pierre had, as usual, arranged the transportation, but thankfully the usual Renault safari bus was substituted for a more comfortable and capable Toyota 4x4 double cab pick-up truck. The journey to Bertoua commenced on the 18th, but soon ran into trouble when the clutch cable snapped. Pierre and the driver managed to make a temporary repair and were able to nurse the vehicle to Bertoua. The first leg of the journey was completed safely. After a few hours' stop, the team pressed on and made it to Yokadouma in the late evening, thanks to the dry, hard packed road and the faster, more comfortable Toyota pickup. The team made it to Moloundou on the 19th of January, which was a record compared to the previous journeys undertaken—three days or more to reach Moloundou.

Prior to leaving Portland, Milt had send Pierre a 12-foot inflatable Zodiac boat with a 20hp engine for river work. Everything had arrived in Cameroon safely, although the courier company and Cameroon Customs expected the recipient of such expensive goods to pay a small fortune for them. After purchasing gas and other supplies, the team headed for a World Wildlife Fund camp located at Dongou near the Dja River. The camp was well appointed with electricity, concrete buildings with tin roofs, guest accommodations, and even wireless Internet and a satellite phone! After an overnight stay at the camp, the team prepared to leave early the following morning, and headed north on the Dja in the inflatable boat.

Pierre had traveled this stretch of the river before and had interviewed eyewitnesses who had seen a *la'kila-bembe* in the past few months. The witnesses were all fishermen and river traders who spent a good deal of their time in their canoes, and hailed from several villages along the Dja. Pete had brought along a binder with detailed satellite images of the river system, kindly provided by atmospheric scientist Ed Holroyd, who worked for the Bureau of Reclamation and is an adjunct professor at the University of Denver where he teaches classes on remote sensing and image processing. His assistance in our research was deeply appreciated by all. As the light began to fade, Pierre indicated that the team should land, as there were no villages nearby to stay the night. He was concerned that they were on the Congo side of the river, but there was no other spot available. Moments after landing, Pierre found a fire pit already prepared with kindling wood ready to burn! He had no previous knowledge of this location, but the whole team was thankful for it, as the darkness of the night was descended rapidly, leaving only the tasks of erecting their tents and preparing an evening meal. So far, the trip had gone remarkably smoothly with no further delays or mishaps. But in Africa, things could go wrong very quickly and even the most thorough precautions were sometimes rendered useless if disaster struck without warning. But now it was time to eat, sleep, and prepare for the morrow. What would it bring?

Pierre speculated that going on the latest reports, the animals were still located in specific places, though far apart from one another, but he was convinced that a small waterfall north on the Dja would be the place to look based on the most recent sighting. The team stopped at six villages on their journey on the Dja, during which time Pierre made some discreet enquiries among the people concerning *la'kila-bembe*. As usual, most people had heard of the animal, but it was most often seen by fishermen and others who spent time on or near the river, such as cocoa farmers who used the river extensively to transport their harvest to Moloundou. The collective fear of the animal was again very evident. The people simply did not like *la'kila-bembes* and went out of their way to avoid any contact with them. Rather than use the illustrations in the binder, witnesses were asked to either draw or describe the animals that they saw in the river. As with all previous eyewitness accounts, the descriptions remained consistent. All six eyewitnesses described an animal with a body sometimes larger than an elephant, a long thin neck ending in a distinctive snakelike head, a long flexible tail, armored skin, and sometimes a series of dermal spikes running the length of the neck, back, and tail.

Upon further questioning, it became obvious that none of the witnesses knew one another (unless they were from the same village), as they lived in places on the river up to 100 miles apart. It is worth noting that the Bantu villagers do not speak the same language as the Baka people, and therefore they referred to the animal as *mokele-mbembe*, rather than *la'kila-bembe* as in the Baka language.

The Zodiac boat and engine performed flawlessly, transporting the team 110 miles upriver to the locations where the animal had been seen, including the "waterfall" which was in actual fact just a series of large rapids. During the river trip, Pete noted at least seven sealed cave locations that he pinpointed and recorded with his GPS unit. The team expected the animals to be in their caves by this time, and the additional data that had been gathered would be valuable in not just planning a new expedition, but to develop new strategies to locate and film one of the animals. It was time to plan a return to Africa with a renewed effort to finally capture our quarry on digital film.

Milt, Pete, Rob, and Pierre finally left the bush on January 27th. The two-day journey back to Yaounde was completed without any dramatic incidents and everyone returned safely home to the USA by February 4th. Although the team did not observe a *la'kila-bembe*, they confirmed that the animals are still active in the region and pinpointed several promising locations along the river that were worthy of further research. With an expedition-friendly government, helpful native people, better timing, and improved night-vision and sonar technology, a breakthrough was now only a matter of time.

> There is always something new coming out of Africa.
>
> —Aristotle

10

THE MONSTERQUEST HUNT

By December 2008, I was working on the possibility of another expedition to Cameroon, when I received a telephone call literally out of the blue from Will Yates, senior producer with Whitewolf Entertainment, a film company based in Minneapolis, Minnesota. Will explained that Whitewolf was responsible for filming and producing some episodes of MonsterQuest, a popular television series dedicated to cryptozoological mysteries and aired by the History Channel. Will explained that Whitewolf was filming for series three, and that he would like to commit one of the broadcasts to *mokele-mbembe*. We discussed this briefly, but the time frame for filming an episode on *mokele-mbembe* was limited to a few months early in 2009. At first I was reluctant to become involved in the project, as the dry season in Cameroon would be in full swing, thus the animals in question would be holed up in their hibernating caves. Furthermore, the limited time for travel and filming, a mere ten days, was clearly insufficient if Whitewolf wanted to fly a film crew out to Africa, travel overland to the border with the Congo, film for three or four solid days on the river system, obtain additional film of the dense African forest, interview eyewitnesses, then travel back to Yaounde and recover for 24 hours prior to departure from the country. I expressed my doubts to Will Yates, and felt that given the lateness of the filming and the short amount of time he could commit to such a project, he would be better off pursuing any film project relating to *mokele-mbembe* closer to November 2009, when the wet season was drawing to a close. Otherwise, it was unlikely that I would want to become involved.

A few days later I received another phone call from Jared Christie, a highly experienced field producer who had been involved in the filming and editing of a number of MonsterQuest episodes. Jared explained that going to Africa a full year later was out of the question, and as he was already committed to producing an episode of MonsterQuest on *mokele-mbembe*, he would still proceed with or without my participation. Beyond the fact that it was highly unlikely that a specimen of *mokele-mbembe* would be observed in the rivers or swamps after February or before September, my concern was also focused on whether or not the documentary would be factually accurate. My telephone conversations and emails between Jared and Will eventually progressed into actually planning a trip to Africa, and with that, a basic outline was hatched to assist Whitewolf in their desire to produce a one-hour documentary on *mokele-mbembe*. The ground preparation in any host nation is essential for all expeditions, and thankfully Pierre Sima was still able to assist us, in spite of being kept very busy between running his plantation and doing translation work for missionaries. The next problem was who would be available to come with me?

The History Channel is an American television channel, and uses various American film production companies to film and edit various episodes or shows for the MonsterQuest series. I felt that it was only appropriate to have an American researcher as my teammate, and drew up a list of possible fellow explorers who would be available to come. Scott Norman had sadly passed away in February 2008, Dave Woetzel was currently engaged in another MonsterQuest film shoot in Papua New Guinea, and both Milt Marcy and Peter Beach were simply too busy to go. Rob Mullin, however, was available, and after conferring with his employers, secured sufficient time off to participate in the project. The following weeks were spent on various tasks and preparations, including drafting up a reasonable operating budget, purchasing various items of equipment, clothing, and footwear, including the all important first aid kit. Pierre arranged for all accommodations and transportation in Cameroon and prepared to meet us at the airport. After receiving and signing our contracts with Whitewolf Productions, Rob and I sent off for our visas while our flights to Cameroon via Swiss Air were arranged. I would meet up with Rob, Jared Christie, and Steve Plummer (cameraman) in Zurich. According to the travel arrangements, I would have to claim my two suitcases in Zurich and re-book them through to Yaounde.

On March 3rd at 7:00 a.m., I said goodbye to my wife and sons before leaving my home to catch an early flight to Montreal via Air Canada. I was bound for Africa once more. Later that day, the connecting flight to Zurich went without a hitch as the efficient and friendly service on Swiss Air seemingly made the journey all too short. Although most airline meals seem to have shrunk in proportion to the increased ticket prices, the bonus of international air travel was that you could always catch up with the latest movies on the newer fleet of Airbus and Boeing aircraft with their state-of-the-art entertainment systems.

My flight finally touched down in Zurich after fourteen hours of traveling. Although jet lag had not yet set in, I decided to find a private or single disabled toilet where I could shave, brush my teeth, and generally freshen up without having to jostle for space with my fellow travelers in the larger, shared toilets. After checking with the Swiss information desk, I was relieved to find that my two suitcases had been booked all the way through to Yaounde and that I would not have to claim them in Zurich and go through two or three different security checks again. Rob and the film crew had not yet landed in Zurich, and with little else to do, I decided to make my way to the still empty gate where our connecting flight to Yaounde would be leaving in three hours. Rob, Jared, and Steve would no doubt show up later. With jet lag slowly beginning to catch up with me, I slouched in the sparse seat at the still empty departure gate, placed my wide-brimmed tropical hat over my face, and slipped into a peaceful doze, which didn't last long. A familiar voice jolted me back to consciousness:

"I must have walked past you a half-dozen times before I realized it was you."

It was Rob Mullin, looking much the same as he did six years earlier and still dressed like G. I. Joe. We embraced, moved over to a small café area, and caught up with the missing years. An hour later, Steve Plummer and Jared Christie arrived at the gate. The team was complete. We were ready for the expedition.

There were fewer Africans on the Younde flight out of Zurich than there were on the Air France flight from Paris. The flight was smooth, the service was friendly, and a delighted Pierre Sima waved to us excitedly as we lined up at the security desk to hand over our passports and yellow fever certificates. A security official, Herman Madonda, suddenly appeared out of nowhere and demanded our passports. He intently examined the American passports but ignored my British passport. I noticed this whenever I was passing through Africa. Perhaps it was because my passport had "European Union" stamped on the cover. Or perhaps it was because Cameroon was once partly ruled by the British. Either way, I was grateful for the lack of attention on this occasion. Concerned that our suitcases and film equipment would be searched by smiling customs officials, eager to see what they could confiscate from us and other passengers, I need not have worried. Pierre possessed a document issued by the Minister for Science and Research, Dr. Madeleine Tchuinté, that guaranteed our smooth passage out of the airport and into the humid night. The beaming smiles of the ever-eager customs officers instantly turned into the gloomiest expressions I had ever seen in Africa, which made it almost impossible for me to suppress my own very obvious smirk. "They'll remember me for that," I thought.

Yaounde, the city of many hills, had not changed a jot, with exception of a new public park near the hotel, and one or two new buildings under construction. The Mercure Hotel had not changed much either, with Rob and I being given Room number 223, the same room Brian Sass and I had shared in 2003! After convening in the restaurant for a late dinner, we were seated at table number 9, the same table that Brian and I were given before. I hoped it was a good sign.

Although Pierre had arranged to collect our government permits, he still needed the minister's signature. Unfortunately, she was out of the country, but sent word that the expedition should proceed to Bertoua where the Governor, Abakar Ahamad, was instructed to affix the magic signature and stamp to officially sanction our trip. It was not good news, as this might lead to further complications that we did not need on a short, two-week trip.

On our first day in Yaounde, I reminded Pierre that we needed to stop at his plantation in Demako to pick up the tents and other camping equipment. It was then that Pierre decided to break the news that the tents had been eaten and damaged beyond repair by termites, and that our camping tables and chairs were also damaged beyond any further use. Our top-of-the-range gear that was supposed to last for 20 years had lasted less than nine. The best part of our remaining day was spent visiting various stores in Yaounde trying to find a tent. Camping gear could easily be purchased in Douala, where most of the ex-patriot French community lived, but finding a seemingly simple item in Yaounde was next to impossible, and even if we were able to find one, it would have been three times more expensive than a similar item in North America. It was days like this that made me yearn for the convenience of Wal-Mart. After several hours of trudging around Yaounde, we decided to give up our search, went to a bank to change our US dollars into CFA, and hoped that the missionaries living in Demako would have a tent to spare. With the first day almost completely gone, we retreated to our hotel where we relaxed, caught up on our sleep, and prepared to purchase additional supplies for the trip to Bertoua the following morning.

Thursday, March 5th, was a more productive day. We purchased bottled water and a few toiletry items at a local supermarket and I acquired a pair of sneakers at a local street market for use during our road trip. During the exchange of money for my new footwear, widespread panic suddenly broke out as three police vans suddenly arrived to arrest a number of street vendors who did not possess a license to trade legally in public. Not wanting to be arrested for purchasing goods from a crooked vendor, Pierre, Rob, and I quickly walked away through the panicking crowds and ducked out of sight into an electrical store.

Finally, our transportation arrived at the hotel to take us to Bertoua. By noon, we loaded our Toyota minibus and headed out of town. The journey to Bertoua was comfortable and without any further dramatics. The government had made an effort to pave the roads between the major towns, and road crews were hard at work along the route. We made it to Dimako just before sundown, and Reda Anderton and her children were there to greet us, along with a young missionary couple, Barry and Desma Abbott. We all breathed a sigh of relief when Barry confirmed that he owned a tent big enough to sleep four people and was happy to loan it to us. As the light completely faded and the mosquitoes were out in full force, we bade farewell to our friends, headed into town, and checked into the Manza Hotel. The rooms smelled musty, and the air conditioning didn't work. Things improved when we asked for another room. Steve and Jared were given a room with an adjoining suite, allowing them ample room to store their equipment for the night. Tired and still dirty after the journey, we all decided to convene to the hotel restaurant for dinner before retiring for the night. Pierre went home and returned with his wife, Lydie, who was as beautiful and as elegant as ever—quite a contrast to us disheveled and grimy travelers. Dinner was the usual chicken with bread, or fried potatoes, which we nevertheless enjoyed, washed down with cold bottles of orange Fanta. A cool shower and a restful night's sleep in our air-conditioned rooms went a long way to rebuilding our strength for the next day—and the forthcoming journey to Youkaduma.

Even the best-planned expeditions to Africa are rarely without problems. First it was the lack of available tents, and now for the second bombshell. Pierre had traveled into town before breakfast to meet with the Governor of Bertoua in order to secure his signature on our government papers. However, the Governor had decided to attend a special open day at a local junior high school, and was simply not available to assist us, knowing full well that we would be in town. Keeping the white man waiting for hours on end with the most trivial of reasons seemed to be an African pastime. Finally, we decided to press on to Youkaduma to see if the *Suprefe* would be willing to sign the document for us. Apart from the bone-jarring ride, courtesy of the standard Renault safari bus, there were surprisingly few police or military road-blocks on this trip, and most of them waved us through with little fuss. On the occasion that we were stopped for a passport check, my British passport with the magic "European Union" stamp was quickly dismissed, while my companions had their American passports scrutinized intently.

We finally arrived in Youkaduma at 9:30 p.m. and found that the Elephant Hotel was under new management, and the electricity supply had been re-connected. The ceiling fans in our rooms were even working, and dinner was the usual fish or omelet with bread, washed down with bottles of cold Coca-Cola. If dinner was predictable, I knew instinctively what breakfast would consist of. True to the limitations of Cameroon's bush hotels, we sat down to yet another round of omelets with bread, mango-flavored jam, and coffee with powdered milk. Pierre was ready to head into town to visit the *Suprefe*, but needed a gift to present to him. As we did not bring the usual bag of goodies to hand out to corrupt police officers and town dignitaries, I reluctantly parted with my new Grundig emergency multi-purpose radio, which I had hoped to keep for the duration of the trip. But we needed that signature and Pierre took the prized item into town, returning an hour later with a broad smile and the all-important signature and stamp on our documents. After loading up our Renault, we headed into town to purchase some thin foam sleeping mattresses before continuing on the next leg of our journey to Moloundou. Barely five minutes into our journey, our safari bus careened wildly into a ditch at the side of the road and threatened to topple over. A rainstorm the night before had made the road treacherously slippery, forcing us to quickly abandon our vehicle for the relative safety of a small shelter to assess the situation.

In spite of our best efforts to push and pull our vehicle out of the muddy ditch, valiantly aided by several local young men, we were forced to abandon our efforts to continue the journey. It almost spelled the end of the trip, as Rob and I worriedly discussed the situation while Steve filmed another effort to get the bus out of the ditch. Finally, the vehicle made it back onto the road, where we left it for the rest of the day. Reluctantly, we all headed back to the hotel, courtesy of a local man who owned a Nissan car. If another rainstorm hit again that night, the expedition would be well and truly over.

After a second night at the Elephant with a steak dinner, fries, and no more rain, we ate breakfast and purchased a bag of fried egg sandwiches for the road. The remaining leg of our journey was slow going, but we finally reaching the frontier town of Moloundou at 4:30 p.m. on Sunday, March 8th. After a one-hour break to recover from the bone-jarring ride and grab a cold drink, we once again climbed in our vehicle and headed for the river. We would reach Langoue before dark.

The ferry was operating normally as we alighted from our vehicle to allow our driver to drive onto the platform. Instead of moving onto the ferry, he backed away, switched off the engine and refused to go any further. In spite of Pierre's protests, the driver simply refused to budge. He sat on the ground and either would not or could not take the safari bus onto the ferry. Tired, frustrated, and with our tempers beginning to fray, Rob and I became more impatient by the minute with the driver's stubbornness. We had come this far and were in sight of our goal.

No amount of protesting would change the driver's mind. He was not going to take the wagon across the river and that was it. Either he was afraid of losing the remaining daylight completely and would not be able to re-cross the river back to town after dark, or he simply did not want to spend the night in a village with "inferior" pygmies. We suspected the latter. Reluctantly, we re-boarded the vehicle and drove back to town, where Pierre quickly located a driver with a large blue dump truck. Once our equipment had been loaded onto the back of the truck, we climbed in the back and enjoyed a hairy roller-coaster ride back to the river and across the Boumba, where we finally reached Langoue just as darkness closed in.

Trying to erect a tent in the dark was not my favorite recreational activity, particularly in the middle of Africa where mosquitoes seemed to relish the soft, delectable skin of the white man. Pierre took up residence in a Baka hut where there was enough room for him to sleep and store our electronic equipment. Rob and I cleaned up a little with the pack of moist wipes that he had brought along. After a modest dinner of leftover sandwiches, we settled down for the night in our loaned tent, which was just big enough to accommodate four delicate white men. However, the first night in a truly jungle environment does not allow one to sleep easily, and I doubt if any of us had more than five hours of actual sleep before arising again at seven a.m.

The first day in Langoue was spent getting our electronic equipment unpacked before heading for the river. Pierre had arranged for the rental of a large wooden canoe with a small outboard engine, and had sent out word to the surrounding villages for any eyewitness who had seen and observed the *la'kila-bembe* to present themselves to the village. As the canoe arrived on time, we headed downstream to the Forbidden Zone where Pierre gave a step-by-step recollection of the large, elephant-like footprints that covered the muddy bank of Swamp island, which he interpreted as belonging to three *la'kila-bembes*, two adults and one very young juvenile. The animals, however, had long gone. The hibernating caves had collapsed inwards, leaving very little trace of them ever having been there. A couple of young men were busy building a wooden jetty for their canoes on the Congo side of the river, and a sandy bank with a small lagoon was located on the northern end of the island where a young couple from Mali were camped. They were of the Bambara tribe who speak the Bamana language, a Bantu tongue related to Swahili and Zulu. They had never heard of our elusive animal, and while our illustrations of some African animals were familiar to them, all our dinosaur illustrations drew a complete blank, as did several different native names for the animal. This confirmed our theory that the animals would not be known to the tribal groups living north of Maroua, the capital town of the Far North Province of the country.

As our first day progressed, Steve filmed Rob placing two MonsterQuest game cameras on game trails along the Dja River, then filmed myself examining a suspected hibernating cave with a air vent, to see if I could determine if a slumbering dinosaur was in residence. By 3:00 p.m., we all had quite enough of the baking heat and headed back to Langoue for an early dinner.

During our absence, the villagers built a very pleasant shelter for us, complete with two chairs, a bench, and a table, for which we were very grateful indeed. Now we could relax and enjoy Pierre's cooking in style! Although it was important to retire early (around 10:00 p.m.) in order to get sufficient rest and sleep prior to a full day's filming, a number of people arrived at the village before sunset, and spent a good portion of the night singing, beating on drums, and arguing. One could never rely on getting enough sleep in an African village!

It was Thursday, March 10th. As usual, Pierre was up at the crack of dawn to prepare a breakfast of fried eggs, fried plantains, bread, and coffee. Word had quickly spread in the area concerning our arrival, and several eyewitnesses presented themselves to us. Pierre had emphasized that he wanted only witnesses who had actually seen the animal, rather than stories passed down through two or three generations. Apart from some local witnesses who were already known to us, such as the plantation owner Norbert Nanga, five others came from villages north on the Dja whom we had never met, nor had they been interviewed or questioned before on the *la'kila-bembe* or any other mystery animals that reputedly inhabit the area. Four of the witnesses were members of the Bagando tribe that inhabits an area of south-east Cameroon, which is now part of the Lobeke National Park. However, the Bagando or Buganda people are also the largest tribe in Uganda, with smaller groups inhabiting Cameroon and Tanzania. Their language is Luganda, the singular form being Muganda, which is part of the Niger-Congo family of languages. Luganda is also tonal, meaning that some words, as with the Baka language, are differentiated by pitch, which is important in the gathering of information on the mystery animals that we are pursuing. Our first and oldest witness was an 86-year-old Bugando widow by the name of Therese Dadjal.

Still physically active and clear-minded, this impressive octogenarian recalled how she saw the animal twice in the Dja River when traveling with her late husband downstream towards Congo.

What interested me about this report is, not only did Therese give us clear and concise details of the animal's description, including the long thin neck, lizard-like head and huge bulky body, but she referred to the animal as *m'koo-m'bemboo*, which is undoubtedly the same as the *la'kila-bembe* and the *mokele-mbembe*. Therese explained that she and her late husband had seen the animals on two occasions spanning a period of 20 years. Typically, they observed the animals when they were browsing on the leaves of trees that grew along the river's edge. Both experiences frightened her, due to the animals' size (bigger than most elephants) and their ferocity when attacking canoes that ventures to close to them.

The next two witnesses, Edino Fermnana and Riayouko Djelf-Aurel, both observed a very large male *m'koo-m'bemboo* in the Dja River, almost completely out of the water. We were taken quite by surprise when both young men stated that the animal possessed a sort of air sac similar to a bullfrog, which it inflated to make very loud bellowing vocalizations. Indeed, Edino mimicked the sound, which was remarkably similar to the audio recording made at Lake Tele in 1981 by the African-American explorer, Herman Regusters. This at least gave some credibility to the occasional reports concerning *mokele-mbembe* vocalizations. After their interview in front of the camera, and without referring to our binder with illustrations of living and extinct animals, both men drew a startling image of the animal they observed, including a neck longer than the tail, dermal spikes, a large bulky body, armored skin and powerful limbs ending in distinctive clawed toes. As with Therese, they had never been questioned before about these animals.

Our final witness was a local Baka fisherman by the name of Bernard Nawouya, who had seen the animals infrequently over the years, but came too close for comfort to a large male *la'kila-bembe* in December 2003—only one month after Brian Sass and I left Lenguoe after discovering *la'kila-bembe* activity in the Forbidden Zone. Bernard had arrived in the area and decided to place his fishing nets in the channel where the caves were located, unaware of the taboo nature of the location. As he busied himself by placing his net across the channel, a huge upswell of water rocked his canoe violently. Startled by this sudden event, Bernard turned around just in time to see a very large *la'kila-bembe* closing in on his canoe. Its long neck towered above him as it moved menacingly towards him. A series of dermal spikes ran the length of it neck down to the back of its bulky body, which was covered with tight, scale-like armored skin. Its lizard-like head with cold, reptilian eyes and gaping jaws terrified the diminutive Baka as he grabbed his paddle and made haste with all his strength, exiting the channel and into the main body of the river. Bernard literally trembled as he recalled his encounter with what was obviously a large male. Going by the discovery that Brian, Pete, and Pierre had made regarding multiple footprints on Swamp Island in 2004, it is clear that the Alpha male, as we shall call it, was merely protecting its territory, which included a feeding area and caves, from intruders, as a female and a calf also inhabited the area at that time.

After the interviews were finished, we headed north on the Boumba River to test our sonar unit and fish camera. As we were already into the dry season, the river was quite shallow in many places, recording a depth of between five to fifteen feet. At the peak of the wet season, one can expect to record depths of up to 40 feet in places. It wasn't until we drifted south into the Ngoko River, then north on the Dja that the river deepened up to fifteen feet with more interesting results. At first we recorded only a number of fish, some quite large, before our sonar unit started picking up some very odd targets indeed. From the Ngoko River we headed north on the Dja, then drifted quietly south with the current. At first there was very little that caught our attention, until two or three sizeable "targets" were picked up on the bottom of the river. Some of these were undoubtedly sunken trees, while others were less clear. Small schools of fish passed beneath us, as did some small crocodiles. Rob operated our sonar unit, a simple but very effective Eagle Cuda 300 Sonar Fish Finder, while I worked with a Fish TV underwater viewing system, which was almost useless in the murky, chocolate-colored river. The sonar performed flawlessly, making contact with fish and crocodiles, including some very odd serpent-like profiles that appeared to be swimming upriver against the current at a depth of between five and seven feet. We were quite mystified at what these might have been, and Pierre merely stated, "There are many strange animals in the Dja."

After three hours of filming our sonar work on the river, we returned to the village for an early dinner before returning to the Dja at 6:00 p.m. for two hours of night filming. We silently drifted south as the pale moon illuminated our canoe, a fleeting shadow gliding on a silver river. Steve filmed us training our flashlights on the darkened banks as we whispered to one another, mortal hunters seeking a phantom prey. As Pierre quietly steered our canoe closer to the water's edge, the tree limbs seemed to reach out like sinewy limbs trying to snatch us away into the inky darkness of the eerie forest. A distant thunderstorm illuminated the approaching clouds, treating us to a cool zephyr as we finally headed back to the safety of dry land and the village. In the wee hours of the night I awoke to the sensation of water dripping onto my face. The storm had caught up with us to seemingly test the water resistance of our tent.

Wednesday, March 11th, was our final day in Langoue. Rob and I slept in until 8:30 a.m., while Steve headed off to the river to film additional footage on the river and surrounding forest. While filming in the Forbidden Zone, a strong current suddenly overturned his canoe, sending Steve and his main camera tumbling. Thankfully, he managed to seize a low-hanging branch and hang on, while at the same time keeping his camera out of the river. Later that morning, Steve and Jared returned to the river to film some re-enactments with *la'kila-bembe* and *m'koombemboo* eyewitnesses. Rob and I walked down to the river at noon to complete a final film shoot with the sonar unit. Again, we made sonar contact with the unidentified serpent-like creatures swimming mid-river and near the bottom. As pythons swim on the surface and the profiles did not match at all known species of freshwater turtle, we speculated that the contacts might have been large eels or perhaps the African lungfish (family Protopteridae), as these can grow over six feet in length. Pierre seemed less convinced and didn't want to spend much more time on the river unless it was absolutely necessary. By 1:00 p.m. we beached our canoe for the last time and filmed our individual interviews at the village. As it was our last day in

Langoue, Pierre headed into town with two villagers to purchase food for dinner and arrange our transportation for the following morning, while we packed up our equipment and clothing. That night we dined on spaghetti with tomato sauce, washed down by hot coffee, with chocolate-coated granola bars for dessert. As we relaxed after dinner, Rob spotted a strange light low on the horizon, moving rapidly from east to west. Jared and Steve were convinced that it was a satellite, but I wasn't so sure. It appeared to be too low on the horizon. I doubted that it was an aircraft, as there was no engine noise or any identification lights that would indicate a commercial flight, and it was too fast for a helicopter. We continued to watch it as it passed by silently and swiftly, until it disappeared. I felt that it was curious that another strange light in the sky would appear when we had a film crew present.

My last hope for a restful night's sleep finally went down the drain when the rains started just after midnight. The downpour lasted until 3:30 a.m., giving me cause to worry about the condition of the roads as we prepared to return to Youkaduma later that morning.

Although we expected our transportation to collect us at 6:30 a.m., a replacement vehicle, a wheezing single-cab Toyota pick-up, arrived two hours later. Finally after arriving at the *Alliance Voyage* depot in Moloundou, we loaded up our safari bus and sat down to omelets and coffee. The roads proved to be solid enough to drive on and we pressed on, reaching Youkaduma and the Elephant Hotel by 6:00 p.m. We immediately ordered dinner, booked into our rooms, showered, and changed. As the water supply had been cut off, we had to wash out of buckets of cold water. I filled my camping shower bag from the bucket, which provided a refreshing shower before dinner. A simple room with a bed and a cool fan really did make up for the rough living conditions we had endured at Langoue. After the usual breakfast of omelets, coffee, and bread, we were on the road by 8:00 a.m. I have lost count how many omelets I have eaten in my years of visiting Africa.

We reached Batouri by 3:00 p.m. and stopped for an hour to relax and enjoy a cold bottle of Coke and some local pastries. Finally, we arrived in Bertoua just after 6:00 p.m., and decided to avoid the Manza Hotel with its stale-smelling rooms and dysfunctional air conditioning. Pierre directed our driver to the Hotel Christiana, a gleaming white-tiled palace with smartly attired staff. Our rooms were neat and clean, with comfortable beds, television sets that worked, and adjoining bathrooms with hot water and fresh towels. The Manza was like a derelict military prison compared to the Christiana. This is where we would stay in future! Pierre arrived at the hotel at 8:10 pm to take us to his home, where we were welcomed by Lydie, who had prepared a fabulous dinner for us on the verandah. It truly was a feast fit for a king. Later, Pierre showed us a Baka training center that he was building behind his home. Only lack of funding was delaying the completion of the impressive structure. I hoped we could help him finding the additional funding he needed to finish this important project.

Our final leg of the long journey back to Yaounde was not without its delays. An impressive Toyota coach arrived at the Christiana to collect us after a breakfast of omelets (what else?), bread with jam, and coffee so strong I was almost bouncing off the walls.

The coach, however was not the one that Pierre had selected, and sure enough, we ran into engine trouble before we even left town. We returned to the depot and changed vehicles, which delayed our departure for yet another hour. Nevertheless, with the improved road conditions to Yaounde, we made good time and arrived back in the capital city by 4:30 p.m. My hopes of getting at least some clothes freshly laundered at the Mercure quickly evaporated, as the laundry service was not available on weekends. Thanks to Lydie, I was able to hand over some clean clothes to housekeeping for ironing. That evening, we walked to the *Dolce Vita* café and enjoyed hamburgers, fries, and ice cream. Pierre, Rob, and I returned to the hotel for coffee before retiring, while the ever adventurous Steve and Jared detoured to the Hilton Hotel to check out the night life, only to be turned away at the door because they were both wearing shorts.

On Sunday, March 15th, we visited a local market to purchase a few souvenirs, packed our luggage for the flight home, and later visited a splendid Chinese restaurant for dinner. The family who owned the restaurant were from Taiwan and did not see many North Americans. Even though the restaurant was closed for the afternoon, they greeted us warmly and treated us like royalty. On the way to the airport, we shared our overcrowded mini-bus with two hulking big-game hunters from Austria who talked lovingly of the beautifully crafted Winchester magnum hunting rifles that they had brought to Africa to blow the wildlife to bits with.

As we confirmed our flights at the check-in counter, an officious female customs official arrived and demanded that I open my suitcase. As I was fumbling with the tiny keys to open the equally tiny padlocks, Pierre appeared and waved our government documents under her nose. We were engaged in important scientific research and were not to be bothered with such trifles. Miss Congeniality took the huff and stomped off, leaving myself and an indignant Pierre to hustle my suitcases through the Swiss check-in desk. It was always sad to say goodbye to Pierre. He is a rock, a friend, and a brother who can be relied upon under any circumstances. I hoped we would all be reunited again soon.

With window seats for all four of us, our flight eventually left Africa far behind as we reached the European continent. At Zurich, we collected our luggage and went in search of our connecting flights. With Steve, Jared, and Rob having to dash to the next gate, I discovered that I had to take a Lufthansa flight to Frankfurt. We were all rushed and did not have time to regroup and say our goodbyes. With the connecting flight to Frankfurt, then a direct flight to Calgary that whisked me home safely, our adventure was already over. Jared and Steve were a terrific team to work with. We had all gotten along quite well, and I hope that one day we will be able to work together again.

There is no doubt that *mokele-mbembes* are still very much alive and active in the river systems that intersect between Cameroon and the Republic of the Congo. Timing is crucial. But we are getting closer to our goal of finally filming one of these remarkable behemoths. I hope that our return to Africa will not be too far away. We have the will and the determination to follow our dream to the end. But time waits for no man, and is slowly running out for us all.

> When you have eliminated the impossible, whatever remains, however improbable, must be the truth.
>
> —Sherlock Holmes

11

MOKELE-MBEMBE—A LIVING DINOSAUR?

AN AFRICAN MYTH?

I am constantly amused by the armchair skeptic who in all probability has never set foot in Africa, yet remains steadfast in his opinion that *mokele-mbembe* is a mere myth. Either countless tribal groups spread over a vast geographical area, all with their own unique customs, religious beliefs, and cultural practices, speaking hundreds of different languages and dialects, had collective nightmares for centuries about exactly the same kind of "mythical" animal, or *mokele-mbembe* is living creature. However, it is quite possible that with the shrinking range of the animals and their gradual absence from those areas where they were once well known and frequently observed, the natives may have relegated their memory to almost mythical status. The only other theory put forth to counter the living dinosaur idea was by the late Carl Sagan, the famous cosmologist and author, who proposed that some remnant memory of living dinosaurs had somehow been passed to modern man over the vast eons of time from our alleged ancient mouse-sized ancestors who were contemporary with the *Tyrannosaurus rex*! Unfortunately, Dr. Sagan was unable to explain just how this astonishing mechanism worked. One can argue that the "dragon" legends of large reptilian monsters reported by over 200 cultures worldwide may refer to the last few dinosaurs that have struggled to survive in an ever-changing world. But this idea will be examined another time, in another book!

The various locations where *mokele-mbembes* are most commonly reported comprise the Republic of the Congo (Brazzaville), the Democratic Republic of the Congo (Kinshasa), Gabon, Equatorial Guinea, Cameroon, and the Central African Republic. They are populated (collectively) by over 59,589,000 people, who speak 729 different ethnic and tribal languages. Very few of these groups, particularly those who live in the most remote villages and settlements, have any regular contact with one another. Although the official languages of most modern African states are the

William Rebsamen

mother tongues of their former colonial rulers, such as English, French, and Portuguese, traditional languages are mostly spoken in the more remote and rural areas. In Cameroon, for example, over 279 tribal languages are still widely spoken. In the Republic of the Congo, 51 different native tongues are spoken, compared to 242 in the Democratic Republic of the Congo, 40 in Gabon, and 34 in Guinea, with at least one hundred more minor ethnic dialects scattered across the Central African Republic and the Democratic Republic of the Congo. Equatorial Africa is populated by a vast variety of different ethnic and tribal groups, all with their own unique religious beliefs and social structures. If *mokele-mbembe* was a mere religious or cultural phenomenon, it would largely be confined to one particular ethnic group, whether it is the Fang tribe of Gabon, the Mongo tribe of the Democratic Republic of the Congo, or the Bamileke tribe of Cameroon. The very idea that certain tribes are merely making up stories of fearsome river monsters to keep rival ethnic groups out of their territory, groups whose very language, culture, and religious beliefs differ vastly from one another, is a palpable absurdity. In the Likouala region alone, several Bantu languages are active, including Bomitaba, Bekwil, Bomwali, Bonjo, Diboli, Fang, Gbava, Koko, Mbandja, Monzombo, Mpywmo, Ngundi, Ngabaka, and Pomo. Of these, only three are part of the Damawa-Ubangi language family, while the rest are localized Bantu languages with a few that belong to the Niger-Congo language group.

The pygmies, who are as unique from one another as are the various African (Bantu) tribes, are also split into various tribes and cultural groups. For example, the Mbenga pygmies, who inhabit the western region of the Congo Basin, speak Lingala, and differ from the Bi-Baya pygmies of southern Cameroon and northern Congo who speak Baka. The Mbuti pygmies of the Ituri Forest in the Democratic Republic of the Congo speak BaBila, which is again different from that of the Efe tribe, who also live in the Ituri, but speak a Central Sudanic language related to Mangbutu, which is part of a Nilo-Saharan language group. The Kango pygmies, who inhabit the western Ituri Forest, speak Komo, which is distantly related to a Bantu language. The Aka, who inhabit the Central African Republic, speak a similar language to Lingala, but are again quite different from the Twa pygmies who inhabit parts of Rwanda, Burundi, Democratic Republic of the Congo, and Uganda, and speak the Kirundi and Kinyarwanda languages. Once again, the notion that from time to time, differing groups of pygmies from all backgrounds share stories about a giant hippo-killing, plant-eating monster that inhabits the rivers, lakes, and swamps over an area stretching 100,000 square miles, long before the white man arrived on the scene, is highly unlikely in the extreme. However, could it be possible that the eyewitnesses have simply mistaken known African animals for an unknown creature?

Mistaken Identity?

No animal living today matches the description of *mokele-mbembe*. The animal is most commonly described as approximating the size of a hippopotamus or an elephant, with a distinctly sauropod-like appearance. The neck is long and thin, no thicker than a man's thigh, ending in an absurdly small head shaped much like an American football, but with facial features similar to a python. The body is bulbous like a hippo, with four strong legs. The tail is described as being long and thin, tapering to a point. The legs are strong and capable of carrying the animal as it moves around on land, albeit close to the water's edge. The skin may have a smooth appearance, reddish-brown in color, but with mature specimens possessing armored skin similar to a crocodilian. The male of the species, we are informed, possess a series of dermal spikes that run the length of the head, neck, back, and tail. The animal is also said to leave large elephant-like footprints in the mud beside rivers and lakes, but with claw marks that elephants do not leave. The description of *mokele-mbembe* simply cannot be confused with any known living animal. The skeptic asserts that the frightened native observers most likely confused an elephant, moving across the river with its trunk raised in the air, for a dinosaur-like creature possessing a long neck. Another theory put forward suggests that all the *mokele-mbembe* sightings are nothing more than native observations of browsing giraffes. The very best way to insult the highly skilled hunters and trackers of Equatorial Africa is to tell them that they are simply wrong in their observations, and that a swimming pachyderm is all there is to the mystery.

Bantu and pygmy hunters are perfectly familiar with elephants, whether they are on dry land or fording a river. Indeed, elephants are still widely hunted in the Congo Basin countries, where even today in the 21st century, various pygmy tribes still track and hunt elephants with little more than large spears to fell the beasts. Although the giraffe (*Giraffa camelopardalis*) possesses a long neck, its thin legs and short tail are very unlike the powerful limbs and long flexible tail of the *mokele-mbembe*. Furthermore, the murky swamps that are home to *mokele-mbembe* are a most unsuitable habitat for any giraffe, which inhabits the Far North Province bordering Nigeria where the Baka pygmies and Bomwali people do not dwell.

The hippopotamus (*Hippopotamus amphibius*) has also been suggested as a likely culprit for *mokele-mbembe*, as it is semi-aquatic and will sometimes even capsize canoes in the river if provoked. Although a female hippo with a calf can be a very dangerous combination under most circumstances, they are nowhere found where *mokele-mbembes* are allegedly present. Having traveled extensively along several major river systems in the Republic of the Congo and Cameroon, including adjacent swamps and lakes, I have not yet observed a single hippopotamus, even though the conditions were ideal for them. My colleagues and I were repeatedly told that *mokele-mbembes* chase them away.

Eyewitnesses

Almost all eyewitnesses we have met during our expeditions have been trackers, hunters, and fishermen, all of whom have spent a good deal of their lives in the most remote and wild environment that one could imagine. Their descriptions of *mokele-mbembes* vary only in the size and coloration of the animals, or in certain features such as the presence (or absence) of dermal spikes. Almost all eyewitnesses observed the animals when they were browsing, most often in the late afternoon, occasionally at night, and sometimes up to several hours at a time. This has given the witnesses plenty of time to observe the animals closely enough to describe specific physical features, including habits and behavioral patterns that have allowed us to extract accurate zoological details. In the case of others who encountered the animals, they have either observed them for mere seconds, as the animals submerge into the river after being disturbed when browsing, while a few others describe how they were ungraciously dumped into the river when a *mokele-mbembe* surfaced directly under their canoes, breaking the vessels in half. Western observers have included a Knight of the Realm, an English aristocratic big game hunter, a colonial governor, two British engineers, several British Army personnel, and two more big game hunters, one French and the other South African. And these are only the ones that we know of. Occasional reports have also been forthcoming from Congolese river boat captains, indigenous pastors on their travels, and Colonel Pascal Mouassiposso, the former military governor of Brazzaville, who observed *mokele-mbembes* in the Likuoala-aux-Herbes River on two occasions. These individuals are sober, level-headed people who were not looking for "dinosaurs," nor did they give them a second thought. But, they nevertheless encountered such—or at least animals that looked like them—when they least expected it. Their own down-to-earth reports only strengthen the "narratives of the natives," and considerably weakens the skeptic's assertions that hippos, elephants, and giraffes are all mistaken from time to time for a swamp-dwelling, hippo-killing suspected dinosaur by people who have lived, hunted, and fished in the Congo Basin region for countless generations.

Although there are many tribal names for *mokele-mbembe* proper, the description of the animal remains almost perfectly consistent throughout west and central Africa. Let us now briefly examine the locations, tribes, languages, and descriptions of the animal that is so obviously well known in various locations around sub-Sahara Africa.

Geographical Locations

SUDAN: The Nuer people of the White Nile region in southern Sudan refer to the *lau* (speaking a Nilotic language closely related to that of the Dinka and Atwot people). The animal is identical to the *lukwata* of Uganda, and is most likely the same.

UGANDA: The Buganda people who live around Lake Victoria are fearful of the *lukwata* (as it is known in the Luganda language), a long-necked, bulbous-bodied, semi-aquatic animal that will capsize their canoes and sometimes kill the occupants, but without devouring the bodies. The animal is still observed occasionally in the White Nile, which connects to Lake No at the confluence of the Bahr al Jabal and Bahr el Ghazal rivers.

CENTRAL AFRICAN REPUBLIC: The Banziri tribe (Kambata language) call the same animal *songo*, and the Baya tribe refer to the animal as *diba* (derived from a Bantu language related to Songho). The animal is said to inhabit the Brouchouchou and Gounda Rivers in the Makala district. In the Yetoname Division the animal is known as the *bagidui* by the Bozoum people (a Bantu dialect called Bozom, related to the Diabe and Boyali languages) and is described as a long-necked plant-eating river monster with a head like a python, only larger and flat on top. It chases away or kills hippos. The name *morou-ngou* is used by the Gbaya tribe (Gbaya-Bozoum language, an Ubangi River dialect).

ZAMBIA: The Barotse tribe of the Upper Zambezi Basin is familiar with the *isiququmadevu* (Makololo language), which is a large animal with a body bigger than a hippo, four stubby legs, a long neck ending in a small head, and a long flexible tail. The animals have been occasionally observed in the Barotseland swamps, which is part of the Barotse Floodplain.

ZAMBIA (LAKE BANGWUELU): The area, including some of the lake's islands is populated by Bemba people (Chibemba language). They are familiar with a large animal with a body bigger than a hippo, four stubby legs, a long neck ending in a small head, and a long flexible tail. They call the animal the *mbilintu*. The animal is also said to inhabit the Barotseland swamps, which is part of the Barotse Floodplain. Englishman Alan Brignall probably observed a specimen of this animal while on a fishing expedition to the lake in 1954.

GABON: The Fang people live on the upper Ogooué River. They spoke of a long necked river monster to herpetologist James H. Powell Jr, called *n'yamala* (Fang language—related to the Bulu and Ewondo languages of southern Cameroon). The animal is an herbivore that kills hippos and menaces canoes that venture too close. This is probably the same animal as the *amali*, mentioned by Alfred Aloysious Smith (Trader Horn) in his biography, *The Ivory Coast in the Earlies*, (Simon & Shuster, New York, 1927).

DEMOCRATIC REPUBLIC OF THE CONGO (KINSHASA): The Bakongo tribe refers to *m'bokale-muembe* (in the Kikongo language, which is related to Swahili, Shona, and Bembe). The description of the animal is identical to the *mokele-mbembe* of the Republic of the Congo, and is described as a hippo- or elephant-sized herbivore with a long neck, small head, and long flexible tail. The animal spends most of it time in the water and will kill hippos and attack canoes.

REPUBLIC OF THE CONGO (BRAZZAVILLE): Several Bantu tribes inhabit the Likouala region, including the Bakongo, Bomitaba, Bomwali, Bonjo, and the Diboli peoples. The most common language spoken in the Likouala and Sangha regions of the Congo is Lingala, which originated from Bobangi, a Congo River dialect. Lingala has become the *lingua franca* of the people, and the name *mokele-mbembe* (meaning, "one who stops the flow of rivers") is derived from this language. It is the only name known for the animal in Lingala, and does not mean "rainbow" or "eater of palm trees" or any other invention circulating the World Wide Web.

CAMEROON (FAR NORTH PROVINCE): In 1932, Ivan T. Sanderson and Gerald Durrell explored the Manyu River in the north of Cameroon where they encountered an enormous, unknown river monster that the natives called *embulu-embembe*. This location is inhabited by several tribal groups, namely the Diaba, Gidar, Mazangwar, and Fali tribes. The languages spoken in this area include Shuwa-Arabic, Mpaele, and Kano-Katsina-Bororro, which are Niger-Congo languages from which the names *ebulu-embembe* and *m'koo-m'bemboo* originate. The aforementioned tribes are spread from the southern tip of Lake Chad to the eastern border, where seasonally inundated rivers and mangrove swamps cover a large area, providing ample space for any large semi-aquatic herbivore to hide, feed, reproduce, and move around freely, virtually without detection.

CAMEROON (SOUTH PROVINCE): Lake Barombe mbo, where some unidentified British soldiers in the company of A. S. Arrey, a Cameroonian Air Force employee, observed two long-necked animals emerge from the lake in 1948, is situated in the South Province near the town of Kumba. The dominant tribes include Beti-Pahuin peoples, such as the Ewondo, Fang, and Bulu. Their dialects are derived from the Beti language, which originated as part of the Niger-Congo language family, and from which the name of another long-necked monster, the *jago-nini* (meaning, "Giant Diver") comes from. The natives assert that the animals only appear once every 20 years or so, which may co-inside with their suggested breeding cycle, which we shall examine shortly.

CAMEROON (EASTERN PROVINCE): Finally, we arrive in the Eastern Province of the country, which is called the "forgotten province" due to its thin human population, poor roads, and overall lack of infrastructure. The Eastern Province is inhabited by the Baka (BaBinga), Bangantu, Bekwili, Bomwali, Mpo, and Njem tribes. All are Bantu groups with the exception of the Baka or BaBinga peoples, who are pygmies. The main tribal languages of this region are Baka, Bomwali, Bulu, Kol, Mbonga, and Vute. The name *la'kila-bembe* is derived from the Baka language of the upper Dja and Boumba River regions, where Baka fishermen and hunters have observed them in the rivers and swamps for generations. The name *mokele-mbembe* is a more familiar term among the Lingala speakers who live on the border with Cameroon and the Republic of the Congo near the confluence of the Dja, Ngoko, and Sangha Rivers. Although some of the Bantu languages, sub-languages, and dialects are distantly related via the Niger-Congo languages and Congo River dialects, at least four of these are now extinct.

A Shrinking Range?

Although there have been no reports in recent years from locations such as Lake Victoria, Lake Bangweulu, the White Nile of Sudan, or the Ooguie River in Gabon, this does not mean to say that a few stragglers have not remained in these locations, but no expeditions have recently explored these areas to determine if the animals are still present there. However, this is certainly not the case in the Likouala Region of the Congo Republic, or the Uele River near the border with Sudan in the Democratic Republic of the Congo, where reports still trickle in from time to time via missionaries and travelers. The larger rivers such as the Dja, the Boumba, the Likouala aux Herbes, the Sangha, the Oubangui, and the Uele provide ample space and range for *mokele-mbembes* to move from one location to another, with countless smaller rivers, swamp pools, and lakes where the animals can hide for weeks or even months at a time, feeding and possibly engaging in reproduction in remote, tranquil surroundings. Reports from the Sangha River and the Dja River concerning *mokele-mbembe* activity have been solid and consistent since the first "modern" expeditions commenced in 1913 with Frehierr Von Stein zu Lausnitz. Although we cannot say with certainty that *mokele-mbembes* continue to thrive outside the Congo Basin, they are certainly still present in the Likouala Region of the Republic of the Congo in the Oubangui, Sangha, Likouala-aux-Herbes, and Bai Rivers, including the remote and tranquil Lake Tele, and further north into the Dja, Ngoko, and Boumba Rivers of southeastern Cameroon.

General Habitat and Environment

We have now established that *mokele-mbembes* are found where the rivers are deep, or at least those rivers that possess deep pools, adjacent swamps and remote, tranquil lakes. Since the first Mackal expedition of 1979, a great deal of emphasis has been placed on Lake Tele, which I personally feel is a mistake. Even if *mokele-mbembes* were still in the habit of frequenting the lake in search of its food supply, it would only be just one of many locations in and around the Likouala Region where they might be found. It is worth taking a closer look at the locations where the animals have been reported, and how the environment is able to support them.

Lake Tele

Lake Tele is an almost perfect ellipsoidal shaped body of water, eight kilometers wide and three meters deep. Swamps surround the lake over a three kilometer radius, and the flooded forests are barely penetrable, particularly in the wet season. Prior to the white man, the lake was called Bangena, after a prominent Bomitaba chief that ruled the area. The word *tele* is Kongo for "new" or "virgin" lake, and assumed this name after the arrival of the French.

Although several expeditions have managed to penetrate the swamps and forests to explore the lake, guides from the village of Boha have ensured that any exploration of the northern region of the lake is kept to a minimum or avoided entirely. The area is still considered taboo by the Boha villagers since the killing of a *mokele-mbembe* in a *molibo* (water channel) located at the northern end of the lake around 1960. Some informants have also stated that *mokele-mbembes* are always observed at that location, where the *molibos* merge with the swamps, which in turn stretch west to the Bai River. Such an undertaking would be time consuming, due to various obstacles such as fallen trees, log jams, and other hazards, but may still yield good results.

In June 1992, a Franco-Congolese multidisciplinary scientific expedition explored the lake to conduct a number of scientific tests. The water in Lake Tele is 40% filled with an organic silt layer a meter thick, and its hydrological exchanges are almost exclusively vertical with very little lateral contribution from the surrounding swamp. The water is very slightly mineralized but is very rich in organic carbon matter and is very acidic. In addition, the team found a magnetic anomaly of some hundred nano-Tesla from an unidentified magnetic source at a shallow location in the lake's northern half. Further research suggested that pollen, whether ancient or recent, comes mainly from colonizing taxa such as *Macaranga*, and suggests that the forest is slowly filling in the lake.

Today, the area is now part of the Lake Tele Community Reserve, a joint project established by the Wildlife Conservation Fund in partnership with the Brazzaville Government, and covers nearly 4,500 km of wetland habitat. It offers protection to various wildlife populations, including high numbers of western lowland gorillas and over 300 species of birds. Twenty-seven villages are located in or around the reserve, and expeditions seeking to penetrate the region may now require extensive funding to acquire the appropriate documentation before being permitted to explore the area.

THE LIKOUALA REGION

The Likouala Region of borders the Democratic Republic of the Congo and the Central African Republic. The region covers an area of 66,044 kilometers, the current population is around 90,000, and the administrative capital is the town of Impfondo. The dry season is from March to July, with the remainder of the year being dominated by the rainy season. The general temperature remains fairly steady at 25°C with high humidity. The main rivers where *mokele-mbembes* have been reported regularly include the Likouala-aux-Herbes, the Sangha, and the Bai Rivers, although the animals have also been observed in the Oubangui River north of Impfondo in recent years.

The Sangha River is undoubtedly the key to all future research and is a major tributary of the Congo River, formed by the Mambéré and Kadeï headstreams at Nola in the southwestern Central African Republic. The Sangha flows 140 miles south to Ouesso in the Republic of the Congo, forming part of the border with Cameroon and the Central African Republic. The river then weaves from southeast to southwest, flowing 225 miles to its mouth at the Congo River, south of Bobaka. The Sangha River is navigable by river barge all year below Ouesso and intermittently as far north as Nola. Its southern, swampy region is connected by smaller, divergent streams with the Likouala-aux-Herbes and Oubangui Rivers, which I suspect is a *mokele-mbembe* migration point. Other smaller rivers and lakes are also worth exploring, such as the Motaba River, north of the Oubangui, and as far as the village of Makako, with a mixed population of perhaps 500 half Bantu and half Bayaka pygmies. The Bai River, which connects in the south at the confluence with the Likouala-aux-Herbes River, stretches north and provides access to two small lakes—these, Lake Fouloukou and Lake Tibeke, situated deep in the swamps directly east of Lake Tele, are also certainly worth further investigation.

Other likely locations for *mokele-mbembes* include the Sangha River north of Ouesso, and the Oubangui River north of Impfondo, including the Likouala-aux-Herbes River south of Epena all the way to its confluence with the Bai River. No expedition to my knowledge has yet explored the Bai River and adjacent swamps as far north as the village of Toukoulaka or the remote Lake Djaka located on the Tibeke River, due north of its confluence with the Sangha River at Ikelemba.

In Cameroon, the most recent area of our research, the Dja, Boumba, Ngoko, Boumba, and Lapondji Rivers have all been sources of recent *mokele-mbembe* reports. However, all our research has confirmed that *mokele-mbembes* are far-ranging animals and may be found throughout the region, in particularly the Sangha River north of Ouesso and possibly as far as Nola. A thorough search of the Sangha, Likouala-aux-Herbes, Ndoki, and Bai Rivers may well produce substantial results, and such an undertaking would be well worth pursuing.

Population differences?

The differences in size and coloration of the *mokele-mbembes* reported in the Congo and the *la'kila-bembes* of Cameroon are striking and deserve further analysis.

Mokele-mbembes as Described in the Congo Republic

Height: Up to 12 feet.

Length: Up to 30 feet.

Facial features: Like a snake or lizard.

Color: Reddish-brown.

Distinctive Features: Some specimens display a distinctive frill adorning the top of the head.

Limbs: Four powerful legs, leaves distinctive clawed prints in the mud beside rivers and lakes.

Diet: Variety of leaves, including the fruits of the *Landolphia* vine.

Habitat: Rivers, lakes, swamp pools, including caves submerged in the rivers and deep pools.

Competition: *Mokele-mbembes*, wherever they are found, do not like hippos and will either chase them away or engage them in combat and kill them outright, but without eating the bodies.

Reproduction: Unknown at this time.

Longevity: Unknown at this time

Behavior: *Mokele-mbembes* are known to capsize canoes by spilling the occupants into the water, and kill them by tail lashing and biting, but without eating the bodies. The natives that we interviewed who experienced this alarming behavior but managed to escape from the animals stated that the animals surfaced under their canoes after being disturbed by human net-fishing activities.

Mokele-mbembe or *La'kila-bembe* of Cameroon

Height: Up to 20 feet.

Length: Up to 40+ feet.

Facial features: Like a snake, specifically a python.

Color: Dark gray to reddish-brown, with armored skin similar to a crocodilian.

Distinctive features: Some specimens display a distinctive series of dermal spikes that run the length of the animal's neck, back and tail.

Limbs: Four powerful legs, holding the body directly underneath like an elephant, leaves clawed prints in the mud beside rivers and lakes.

Diet: Variety of leaves, including the fruits of the *Landolphia* vine.

Habitat: Rivers, lakes, swamp pools, including caves submerged in the rivers and deep pools.

Competition: *La'kila-bembe* will either chase away hippos and elephants, or engage them in combat. It is also said to drive away crocodiles from its immediate feeding area.

Reproduction: Unknown at this time, but said to give birth to a single large calf. More research is required to determine the precise means of offspring and reproduction.

Longevity: Unknown at this time

PHYSICAL FEATURES

The eyewitnesses often describe animals that are so big that they have been unable to pass them safely in the river. A few reports reveal encounters at night while the animal was feeding, and at least two independent reports involved the capsizing of boats, which were broken in half by the dermal spikes that adorn the neck, back, and tail of the animals.

The difference between *mokele-mbembes* observed in the Congo and those of Cameroon are striking and thought-provoking. The animals present in the Likouala Region are often small, with a body not much bigger than a hippo. The head height is between eight to ten feet, with a total length of up to thirty feet. An adult hippo can weigh up to four tons and is extremely dangerous if provoked, and a female hippo with a calf is a most dangerous combination under any circumstance. Hippos are easily able to bite a human being in half, and kill more people every year than any other African animal. *Mokele-mbembes* are said to be far more dangerous than even the most aggressive hippo, and regularly chase them away or kill them in combat. As both animals are semi-aquatic herbivores and are thus in competition over habitat, even if they do not share the same food supply, the animosity between the two is understandable. Elephants, too, are said to be at risk when crossing a river too close to the lair of a *mokele-mbembe*, but conflict between both animals are rare in the Likouala. Congolese villagers who observe a *mokele-mbembe* in the river usually turn and paddle quickly away in the opposite direction, or land on the nearest bank and flee on foot. A few fishermen have had the unfortunate experience of being thrown into the water by *mokele-mbembes* surfacing under their vessels after being disturbed by their fishing activities, usually involving nets that have been unwittingly lowered into a deep river or swamp pool where the animal was presumably hiding.

The smaller *mokele-mbembe* of the Congo is said to possess a smooth skin, reddish-brown in color, with a few specimens possessing a comb-like frill adorning the head. A few larger animals are encountered very occasionally, with one enormous specimen (said to be as large as five elephants!) being observed by a family of three from Boha village in the Likouala-aux-Herbes River in 1981.

The larger *mokele-mbembe* or *la'kila-bembe* of Cameroon is said to be at least twice the size of the Congo animal, with a body bigger than an elephant and a head height of at least a giraffe. The largest giraffe can reach 18 feet in height and weigh up to 4,400 lbs. The African male savannah elephant (*Loxodonta africana*) stands up to 12 feet at the shoulder and weighs an average of 12,000 lbs.

If the *la'kila-bembe* is able to reach at least the height of a giraffe with a body approximating the size of an elephant, its body weight would be well in excess of 15,000 lbs, taking its armored body, long neck, and tail into consideration. If our eyewitnesses are reliable, the *la'kila-bembe* can easily kill hippos and elephants by rearing up on its hind legs and striking them with its foreclaws and tail. Such a scene immediately brings to mind the same sort of defensive posture that has been suggested by paleontologists and literally brought to life in movies such as *Jurassic Park*, and the popular television series, *Walking with Dinosaurs*. Needless to say, the eyewitnesses who dwell in small villages and camps along the rivers and swamps do not possess television sets or attend movie theatres.

Armored Skin

Certain physical features of the *la'kila-bembe* are most intriguing, particularly the armored skin and dermal spikes. The fossil record has revealed that some sauropods such as the *Saltasaurus* and the *Laplatasaurus*, both members of the family Titanosauridae and known from the fossil record in South America, were protected by hundreds of small bony deposits (osteoderms) about the size of peas, tightly packed in the skin, with the *Laplatasaurus* possessing a few large oval bony plates about as broad and thick as a person's palm. This bony protection seems to have covered the back and sides of its body and probably gave it a roughened, bumpy appearance. Other sauropods also possessed defensive armor and at least one, *Shunosaurus*, from the fossil record in China, even sported a clubbed tail.

Dermal Spikes

The reported presence of dermal spikes or spines on *la'kila-bembes* proved to be an interesting development in our search for the animal and was quite unexpected. In 1990, dinosaur artist Stephen A. Czerkas found odd fossilized impressions on some rocks in the Howe Quarry in Wyoming, associated with the fossils of nineteen sub-adult specimens of *Diplodocus* that had perished together. The evidence of dermal spines on the animals, some of them *in situ* on the top ridge of the tail, has since changed the way that paleontologists now look at sauropod dinosaurs. The narrow, pointed spines on the midline of the back of some diplodocids were apparently not a part of the backbone and may have resembled the spines or "comb" seen on the back of an iguana or other lizards, only much larger, up to four inches. One year after Czerkas made his fascinating discovery, Argentine paleontologists Leonardo Salgado and José Bonaparte discovered the fossilized remains of *Amargasaurus*, a genus of dicraeosaurid sauropod dinosaur. It was a small sauropod, reaching about 33 feet long, that sported two parallel rows of tall spines down its neck and back, much longer than what has been found on any other sauropod.

During our questioning of the Baka people regarding the physical features of the *la'kila-bembe*, they stated that the male of the species possessed the dermal spikes, but the female (which had a longer neck than the male) did not. We asked our informants to draw images of the animals in the soft earth, which revealed distinctive sauropod-like animals with armored skin and dermal spikes. Prior to our arrival in Cameroon in November 2000, no one had ever questioned the Baka people about these animals since the Von Stein expedition of 1913, nor were they ever shown illustrations of dinosaurs or other presumed extinct animals by white outsiders of any nationality. As they have never been paid or given any reward for providing us with information on a variety of animals, known and unknown, they simply had no motivation for deception.

The presence of a comb-like frill adorning the head of the Congolese *mokele-mbembe* and the larger dermal spikes on the *la'kila-bembe* of Cameroon may suggest a straightforward case of sexual dimorphism within the species itself. Rob Mullin has put forward the theory that the dermal frill of the Congolese *mokele-mbembe* may be a physical characteristic of the juvenile animal that later develops into the larger, more prominent dermal spikes of the adult as they mature. Another theory that could explain the distinctions between the two would involve the presence of two different species of (possible) sauropod dinosaurs in the Congo Basin. This theory, however, would only cause more problems than we are trying to solve, so we will stick to the Mullin Theory for now.

Aquatic Inability?

Another challenge to the living sauropod theory surrounds the now outdated notion that some sauropods, particularly the giants such as *Brachiosaurus* and *Apatosaurus*, spent much of their time in water to support their massive bodies. Most paleontologists generally believe that sauropods were built for life on dry land, unlike old textbook illustrations that depicted them wading in swamps and sometimes entirely submerged and holding their necks high to breathe at the water's surface. When paleontologists examined aspects of a sauropod's probable respiratory anatomy, this raised the problem of their being unable to breathe when fully immersed. The pressure brought to bear upon the soft trachea and submerged lungs of most sauropods could not have facilitated the ability to either hide underwater or breathe while almost completely submerged in water. But the case is not yet completely closed regarding the alleged aquatic inability of all sauropods, as new discoveries about them and other dinosaurs are being made almost daily in fossil digs around the world. After all, there is only so much we can learn about the precise physiology of the dinosaurs from fossilized bones, and *mokele-mbembe's* human observers are absolutely emphatic that the animal is semi-aquatic and can remain submerged in a river or swamp for long periods of time, like a hippopotamus.

Reproduction

Immature *mokele-mbembes* have been rarely observed, unless we accept that the smaller specimens contained within the confines of the Likouala Swamps are juveniles. We have at least four testimonials from different individuals who have observed the animals mating in the river. We are told that the adult *mokele-mbembes* copulate "like dogs," with the male, possessing the shorter neck but with the presence of dermal spines, mounting the female, who has the longer neck with an absence of dermal spines. No one has yet reported finding a nest with eggs that belong to a *mokele-mbembe,* and so the natives assume that the animals give birth to one live calf perhaps only every 20 years or so, which the female raises in a large protective nest constructed within the confines of a shallow swamp close to the river. Only two independent reports have been forthcoming concerning infant *mokele-mbembes*. On both occasions, a calf, smaller than a hippo, was observed with a very large female *mokele-mbembe* in the river. On one occasion in the Dja River, a Baka hunting party encountered a female and a calf on a sandbank in the middle of the river. Both animals hastily retreated into the water when they became aware of the noisy, frightened humans nearby. Alleged nests have been found in shallow swamps, comprising of a robust enclosure similar to a birds nest, up to twenty feet across and made from tree branches and foliage, but long after the animals abandoned the dwelling.

The live birth theory may not entirely rule out *mokele-mbembe* from being a sauropod, but paleontology has thus far revealed that almost all known species of sauropod were egg layers. Paleontologist Robert Bakker has suggested that the large pelvic canals found in sauropods may argue in favor of live births. Many modern viviparous reptiles give birth to live young, such as certain skinks, Jackson's chameleons, South American boas, most vipers, and the common garter snakes. In the case of reptile viviparity, the eggs are retained in the maternal oviducts and hatch in uterus before birth, but whether or not this is the case with *mokele-mbembe* remains a mystery. If *mokele-mbembe* is an egg layer, the eggs could be consumed by large monitor lizards and pythons, while surviving offspring would still be vulnerable to large predators like the Nile crocodile. This may account for the apparent rarity of juvenile specimens that successfully grow to maturity. The animal's apparently slow rate of growth, which is obviously tied in with its reproductive ability, is not unknown in the world of reptiles. The two species of tuatara from New Zealand are the only living survivors of the order Sphenodontia. Tuataras mature very slowly, reaching sexual maturity in one or two decades. Females can lay eggs only once every two to five years, depending on how long it takes for her to accumulate yolk (a very slow process) prior to fertilization. Then, it may take up to nine months to form the soft shell and they are laid—it may be another 16 months until the eggs hatch. This is obviously the slowest reproductive rate in any reptile, but we may speculate that *mokele-mbembe,* if it is a relic species, is in the same boat. Thus, a slow reproductive rate and the dangers faced by juveniles, ensuring that very few reach maturity, may well explain their apparent rarity.

Diet

Mokele-mbembes are said to be entirely herbivorous, and do not consume the flesh of any animal or human that it kills. Not all reptiles are carnivores, of course, and many lizards and chelonians devour vegetation. An herbivorous reptile does not pose a problem for the existence of *mokele-mbembe*, which is said to consume several different kinds of leaves, and the fruits of the *Landolphia* vines, *Landolphia mannii* and *Landolphia owariensis*, which produce apple-like fruits similar in nutritional value to domestic pears. The vast majority of encounters between humans and *mokele-mbembes* occur when the animals are engaged in browsing along the river's edge.

A Forked Tongue?

A few reports suggest that *mokele-mbembe* possesses a long forked tongue similar to snakes and certain lizards. These reptiles smell by using the tip of their tongue to direct airborne particles to their Jacobson's organ—the forked tongue allows them to determine from which direction a smell or odor emanates. It is not known if *mokele-mbembes* might use a forked tongue to seek out its preferred (herbivorous) food supply or to distinguish edible leaves from those that may harm the animal. They have been observed using their elongated tongues to secure the leaves and fruits upon which the animals reputedly thrive. Only a close examination or film of an actual specimen in the act of browsing will settle this question.

Vision and Hearing

Natives who encountered *mokele-mbembes* at night have stated that the animals' eyes possessed a bright yellow glow when reflecting flashlights or gasoline lanterns. This eyeshine suggests good nocturnal eyesight.

Many reptiles are sensitive to movement, and this was particularly demonstrated when Pierre Sima, who possesses extraordinary tracking and hunting skills, silently pointed to a large two-meter-long Nile monitor lizard to me one day as it sunned itself on the bank of the Dja River. The lizard was able to detect my slow, deliberate movement as I attempted to raise my camera and shoot some film—it retreated into the forest literally in a flash. In 1954, when Allen Brignall spotted a long-necked reptilian creature in Lake Bangweulu, a slight movement from him attracted the creature's attention and sent it sinking back into the lake with haste. *Mokele-mbembes* appear to be acutely aware of any sound or movement that may be "foreign" to them, such as the sound of human voices or the abrupt appearance of motorized canoes. In a few cases, the animals have grown so large and formidable that they have ceased to be concerned with canoes or even large crocodiles venturing near.

Seasonal Dormancy

Winter dormancy is known in mammals (as hibernation), reptiles (as brumation), and other animal groups. Dormancy can be prompted by climatic changes other than cooling temperatures, however. Dormancy triggered by a dry season, when food and water are scarce, is called aestivation. Many snakes, lizards, crocodilians, turtles, and tortoises aestivate during dry seasons.

This is certainly the case with *mokele-mbembes*, which remain inside their sealed caves or chambers from February to September, during the dry season when the water levels in the swamps and rivers are at their lowest. There have been a few rare reports of the animals feeding in the dry season, usually near deep pools in the river where they are able to retreat and remain concealed if disturbed.

Size

In February 2004, Brian Sass and Peter Beach discovered that the animal responsible for the deep claw marks in the steep clay banks possessed a single large scalloped claw used primarily for digging. Peter Beach made some excellent plaster casts of the clawed prints, and later took some photographs of a series of large footprints on Swamp Island directly opposite the high bank on the Congo side of the Dja where the sealed caves were located. Upon careful examination of the prints, Pete was able to establish that while the animal stripped the leaves from the branches of the trees to a height of 18 feet, it shifted its weight to its rear legs for a time, as its rear footprints were markedly deeper in the mud than its fore claws. This may indicate that the animal was reaching up to higher branches. The distance between the two front legs was 64 inches, with a stride between six and seven feet. The animal possessed five toes on each front foot, two of which had narrow pointed claws and one having a large spoon-shaped claw between three and four inches, that were apparently used for digging. The approximate ground-to-shoulder height may therefore be about eight feet, possibly up to twelve feet for a mature female specimen.

Conclusions

What are we to make of the collective evidence thus far? The time has come to put the pieces together and carefully consider exactly what kind of animal is at the center of the mystery that has intrigued and puzzled scientists, explorers, and armchair skeptics alike for over 200 years.

Having discussed the question of the most likely identity of *mokele-mbembe* with leading cryptozoologists, including Roy P. Mackal, Karl P.N. Shuker, Peter G. Beach, and others, we draw certain conclusions:

> *Mokele-mbembe* is in all probability a living animal.
> Its general morphology is strikingly similar to a small sauropod dinosaur.
> It possesses dermal spines and armored skin similar to some sauropods known from the fossil record.
> It is a reptile.
> It is semi-aquatic.
> Its diet is herbivorous.
> It is rare.
> It is dangerous.
> It rears its young in large nests constructed in shallow swamps.
> It range has shrunk considerably since 1800, but it is still widespread throughout the Congo Basin countries.

Can we say with absolute certainty that *mokele-mbembe* is without any doubt a living dinosaur? The answer, in all honesty, is "no." The animal may look like a dinosaur, but that does not make it one, although this is the most popular theory regarding the identity of the creature. Crystal-clear digital film evidence that can yield precise zoological data on the animal will help enormously. Of course, the most important task will be to secure an actual specimen of a *mokele-mbembe,* or at least part of one to provide DNA evidence, to finally settle the matter.

I do, however, agree with Roy Mackal regarding his own considerations about *mokele-mbembe*. It is a living animal, and one that simply does not fit the description of any known living creature within the current repertoire of contemporary zoology. In the end, we dare not make any careless or sensational claims regarding the actual identity of the *mokele-mbembe*. That claim we can only make when we have a specimen of the animal to present to the world.

Mokele-mbembe: Ethnological Variants

LOCATION	TRIBE	NAME OF ANIMAL
Cameroon - SW Province	Chamba people	Embulu-embembe
Cameroon - Central Province	Yaunde tribe	Nwe
Cameroon - South province	Beti-Pahuin tribe	Jago-nini
Cameroon – East Province	BaBinga tribe	La'Kila-bembe
Cameroon – East Province	Bagando (sub group)	M'koo-m'bemboo
Central African Republic	Banziri tribe	Songo
Central African Republic	Baya tribe	Diba
Central African Republic	Bozoum tribe	Bagidui
D.R. Congo (Kinshasa)	Bakongo tribe	M'bokale-muembe
P.R. Congo (Brazzaville)	Bakongo/Bomitaba/Bomwali	Mokele-mbembe
Gabon	Fang tribe	N'Yamala
Sudan	Dinka tribe	Lau
Uganda	Bugando tribe	Lukwata
Zambia	Barotse tribe	Isiququmadevu
Zambia	Njumbo tribe	Mbilintu

The Congo Basin

1 Cameroon
2 Central African Republic
3 Republic of the Congo
4 Congo-Brazzaville
5 Gabon
6 Equatorial Guinea

Noteworthy Sighting Locations in Cameroon

1 Barombi Mbo
2 Manyu River
3 Sanaga River
4 Nyong River
5 Dja River
6 Boumba River
7 Ngoko River

Noteworthy Sighting Locations
in Central African Republic

1 Ouham River

Noteworthy Sighting Locations in Gabon

1 Ogoouie River   2 N'gounie River

Noteworthy Sighting Locations in Zambia

1 Lake Bangweulu

Noteworthy Sighting Locations in or near Uganda

1 Lake Victoria   2 Lake Albert
3 Swamps of the White Nile, Southern Sudan

> I can appreciate the opinion of a great many zoologists and paleontologists who believe that the survival of any dinosaur into the present time is improbable. I cannot agree that the idea is impossible, which is what makes the whole question so interesting.
>
> —Roy P. Mackal

12
N'GOUBOU - THE KILLER OF ELEPHANTS

Quite apart from *mokele-mbembe*, the terror of the Congo Basin, another animal is frequently mentioned by the native people—one that is apparently even more dangerous than our suspected dinosaur. Stories describing a large semi-aquatic animal about the size of a hippo, sometimes as big as an elephant, possessing a single horn which it uses to attack and kill elephants, buffalo, and unwary native fishermen, abounded for generations in and around the Likouala Region. At first it was thought that such reports were confined to the swamps located between the Sangha and Oubangui Rivers, but in recent years my colleagues and I have uncovered reports of the animal in Cameroon, where it is well-known, so may be more widespread than we previously thought. Could this much-feared killer be nothing more than an ill-tempered rhinoceros?

Five species of rhino exist in the world today. Two of these are found in Africa: the black rhinoceros (*Diceros bicornis*) and the white rhinoceros (*Ceratotherium simum*). The black rhino boasts four subspecies, of which the most common, the south-central black rhinoceros (*Diceros bicornis minor*), at one time ranged from the savannahs of central Tanzania to northern South Africa. It may have once been found in the southern Democratic Republic of the Congo, but widespread poaching has severely reduced its range. The subspecies is now extirpated from several countries. The western black rhinoceros (*Diceros bicornis longipes*) was once found throughout the savannahs of central-west Africa, with poachers reducing the population to survivors in northern Cameroon. Recent surveys there did not turn up any rhinos, however, so the subspecies is now considered extinct. The white or square-lipped rhinoceros is second only to the elephant as the most powerful land animal in Africa. There are two subspecies of white rhinos. There are approximately 14,500 southern white rhinoceros (*Ceratotherium simum simum*) in the wild, mostly in South Africa. However, the northern subspecies (*Ceratotherium simum cottoni*) is regarded as gravely endangered and may be extinct in the wild. The latest survey in Garamba

National Park (Democratic Republic of the Congo) for the last four known individuals failed to find any sign of them. The remaining three species of rhinoceros include the Indian, Javan, and Sumatran rhinos. All of these live in Asia and are therefore not of immediate interest in our search for the horned terror of Africa.

One of the first individuals to bring to light rumors of a rhinoceros-like creature from our region of interest was Sir Harry Johnston. Johnston, who was a governor of Uganda, and to whom we owe the discovery of the okapi (*Okapia johnstoni*), wrote in his book *Liberia* (1906):

> "The Mandingos of the northern parts of Liberia assert that a Rhinoceros exists in their country. They recognized at once pictures of the common two-horned rhinoceros, and named it "kowuru." This is a very interesting question to be examined into in the future. It seems strange that the rhinoceros should be absent from the open country of West Africa at the back of the forest, since it is found so abundantly in Eastern Sudan, in Eastern, South-western, and South Central Africa. But although its existence is constantly asserted or reported by Arabs, Hausas or Mandingos in the regions of West Central Africa between Lake Chad and the Upper Niger, no vestige of a rhinoceros has ever been sent to Europe or America from those regions. Until therefore some direct evidence of its existence can be laid before us we must consider the question 'not proven.'"

The French hunter Georges Trial reported that he had observed a rhinoceros in the savannah of the lower Oogue River of Gabon in 1931 or 1932:

> "It was a formidable animal, of an extraordinary length, which appeared so inordinately long to me that it most likely was far less than it actually was. With its nose to the ground, it carried its monstrous head low, dominated by two very high nose horns, approximately equal in length and curved towards each other. It gave the impression of being armored, covered with broad rigid grayish plates, separated from each other by clear wrinkles, and arranged like articulations or a bellows. Besides its double defense I clearly saw the small ears on its misshapen head in continual movement, and its massive rump had the small silly tail of a pig, which it shook frantically. The rhinoceros entered open ground, without suspecting my presence in the least, angled across the plain, then moved away peacefully while

uttering small grunts like a satisfied pig." (Trans. from the French, Trial 1955).

Another early account of a mysterious horned animal came from an Englishman, C. G. James, who lived in Africa for 18 years in the early part of the 20th century. He wrote a letter to the *Daily Mail* newspaper, published on December 26th, 1919. In reference to the mystery animal, he penned the following:

> "Sir, I should like to record a common native belief in the existence of a creature supposed to inhabit huge swamps on the borders of the Katanga District of the Belgian Congo—the Bangweulu Mweru, the Kafue swamps. The detailed descriptions of this creature vary, possibly through exaggerations, but they all agree on the following points; It is named the Chipekwe; it is of enormous size; it kills hippopotami (there is no evidence that it eats them, rather the contrary; it inhabits the deep swamps); its spoor is similar to a hippo's in shape; it is armed with one huge tusk of ivory."

Big game hunters, explorers, and a few missionaries also picked up reports of the *chipekwe*, and it wasn't long before the animal was being mentioned in books written by adventurous explorers. In his book, *Far Away Up the Nile*, (London, 1924), John G. Millias wrote:

> "I have met only one practical hunter and man of observation actually believing in the existence of a great beast unknown to science. This is Mr. Dennis Lyall, who has written many books on the game of Central Africa. He is convinced that there is, or was till recently, some large pachyderm, somewhat similar in habits to the hippopotamus, but possessing a horn on the head, which frequents the great marshes and lakes Benguelo, Mweru, and Tanganyika. He calls it a water rhinoceros, and can deduce good evidence for his theory."

Unfortunately, neither Millias nor Lyall, in spite of their love for adventure in Africa, were able to contribute much more to the mystery beyond a few reports that occasionally came their way.

As with *mokele-mbembe*, there are various names attributed to a large, semi-aquatic horned animal that attacks elephants, hippos, and even occasional canoes without provocation. One of the very earliest written accounts of such a creature appeared in the book *Eighteen Years on Lake Bangweulu*, authored by J. E. Hughes in 1933, in which he reported that an animal fitting the description of an *emela-ntouka* (although not referred to by this name) was killed by Wa-Ushi tribesmen along the shores of the Luapula River, which connects Lake Bangweulu to Lake Mweru.

> "It took many of the best hunters the whole day spearing it with their large 'viwingo' harpoons: the same as they use today for the hippo. It is described as having a smooth dark body, without bristles, and armed with a single smooth white horn fixed like the horn of a rhinoceros, but composed of smooth white ivory, highly polished. It is a pity they did not keep it, as I would have given them anything they liked for it."

Hughes also met Robert Young, an official in the British South Africa Company administration in the Chinsali District of northern Zambia. Young was hunting around Lake Bangweulu and shot at a huge horned beast as it surfaced briefly. He did not know if he hit the animal, as it made off into the lake, churning the water like a "screw steamer."

In January 1931, Monsieur Geraud de Glassus, an official with the French colonial administration in Cameroon, informed Lucien Blancou, a game inspector in the Likouala region of the Congo, that there were rhinoceroses present in the dense forested region in the south, which was part of the Batouri subdivision, where he administered at that time. This information intrigued Blancou, who thought that perhaps the animals were remnants of the savannah rhinos. However, shortly after his promotion to chief game warden for the Likouala district of Middle Congo (Congo Brazzaville), he received further verification of an odd forest-dwelling rhino from three other French administrators, Millet, Rolland, and Rozan, all of them keen hunters who were residents in the Epena District. Another European, an engineer in the Forestry Department, named Moriaud, reported the presence of a rhino-like forest animal to Blancou, who began to investigate the reports more closely.

Bernard Heuvelmans translated the following information from Blancou's private notes:

> "The Africans in the north of the Kelle district, especially the pygmies, know a forest animal larger than a buffalo, almost as large as an elephant, but which is not a hippopotamus. Its tracks are only seen at long intervals, but they fear it more than any other dangerous animal. The sketch of its footprint which they drew for M. Millett is

WILLIAM REBSAMEN

that of a rhinoceros. On the other hand, they do not seem to have said that it has a horn, though they have certainly not said that it has not. While M. Millett was at Kelle, in 1950 if I am not mistaken, one of the best known African chiefs in the district came several days march to inform him that 'the beast has reappeared.' ... Around Ouesso the natives talk of a big animal which does have a horn on its nose—though I don't know if it has one or several. They are just afraid of it as the Kelle People."

Blancou concluded:

"Around Epena, Impfondo and Dongou, the presence of a beast which sometimes disembowels elephants is also known, but it does not seem to be so prevalent there now as in the preceding districts.

"A specimen was supposed to have been killed twenty years ago at Dongou, but on the left side of the bank of the Ubangi and in the Belgian Congo."

Native accounts of the animal in Zambia and the Congo describe it as being as large as a buffalo, sometimes as large as a forest elephant, with smooth grayish-brown skin and armed with a large, single nasal horn having the appearance of smooth ivory. The animal is well-known to the inhabitants of Impfondo, Epena, and Dongou, who describe the animal as being quite different from the known species of rhinoceros. It shares a hatred of elephants and hippos with the *mokele-mbembe* proper, but they are two distinctly different animals. There are no reports, however, of any direct conflict between *mokele-mbembe* and the *emela-ntouka*, perhaps because they are solitary animals for the most part, unlike elephants and hippos that tend to herd in fairly large groups and are thus more common. Although the *emela-ntouka* is sometimes described by one or two native or pygmy groups as *mokele-mbembe*, this confusion is due to the fact that some of the informants have never actually seen the animal for themselves, while other groups use the term *mokele-mbembe* in a generic sense to describe any animal that is unfamiliar to them. It should also be noted that the Lingala word for "animal" is *yama,* and this word is used to describe animals of all kinds, known and unknown.

It would be helpful to again examine the different locations where the horned animals have been reported, including the tribal groups that are familiar with this animal, in order to clearly distinguish the *emela-ntouka* from the *mokele-mbembe* proper.

DEMOCRATIC REPUBLIC OF THE CONGO (CONGO-KINSHASA): The four principal tribes (of the nearly 200 tribes populating the country, mostly Bantu) are the Mongo, Kongo, Luba, and Mangbetu-Azande. These do not include the pygmies that inhabit the forested regions. Out of the 242 known languages spoken, the main four are Lingala, Swahili, Kikongo and Tshilub, which are part of the Niger-Congo language group. The name *chipekwe* originates from the Kikongo language, and the description of the animal is consistent with the *emela-ntouka* of neighboring Republic of the Congo or Congo-Brazzaville.

REPUBLIC OF THE CONGO (CONGO-BRAZZAVILLE): A large and ferocious aquatic horned animal known as the *emela-ntouka* (loosely meaning "killer of elephants" in the Lingala language) is said to inhabit the rivers, swamps, and forests of the Likouala swamps. In the Lake Tele area, the Bomitaba people, who number around 9,600 in the Likouala region, refer to the animal as *emia n'touka* in the Mbomitaba language (classified as part of the Niger-Congo, Bantu language groups). The BaKongo people, whose language is part of the Benue-Congo branch of the Niger-Congo languages, refer to the same animal as the *aseke moko*. The Bomwali people, a small Bantu group numbering about 35,000 in the Likouala, also refer to the animal as the *ngamba-namae*, in the Sangasanga language, a Niger-Congo and Atlantic Congo language, which is also spoken to a much lesser degree in Cameroon.

REPUBLIC OF CAMEROON: The Baka or BaBinga people refer to a large aquatic horned animal as *n'goubou* in the Baka language. Although this world is often used to refer to the rhinoceros, the black rhinoceros is now considered extinct in northern Cameroon, and is very rarely observed in the savannah of western Cameroon and the Central African Republic. The name, however, is commonly used by the Baka along the upper Dja and Boumba Rivers to refer specifically to two distinct animals. The first animal is larger than a hippo, sometimes as big as an elephant, and possesses between one and two prominent nasal horns. It makes its home in the rivers and swamps. The second animal is also as large as an elephant and is described as possessing between three and six horns, a distinctive neck frill, and an armored body. It is observed in the savannah region of eastern Cameroon.

ZAMBIA: Early reports connect Wa-Ushi people of eastern Zambia with the killing of a large horned animal in the swamp south of Lake Bangweulu on the Luapula River. The lake is about 45 miles long and covers an area of 3,800 square miles, including the adjacent swamps. Its outlet is the Luapula River, which is a headstream of the Congo. It also has three inhabited islands. The Wa-Ushi or Ba-Ushi people, as they are sometimes known, are a Bantu sub-group belonging to the Bembe tribe that dominates the area. The language spoken by the group is also called Bembe, but is sometimes known as Beembe, Ebembe, Ibembe, or Kibembe, and is a Niger-

Congo Bantu language. Another Bantu language, Chichewa, is the national language of the Republic of Malawi, and is spoken by Bantu groups in Mozambique and Zimbabwe. It is also one of the seven official tribal languages of Zambia, where it is spoken mostly in the Eastern Province. The Chichewa word for rhinoceros is *chipembere*, which is vaguely similar to the name *chipekwe* of the Kikongo language and also refers to a semi-aquatic horned animal thought to inhabit the Luapula River and surrounding swamps.

Range

The presence of an aquatic or amphibious horned animal has been reported from Lake Bangweulu in Zambia, to the Likouala swamps in Congo, and the lower Dja and Boumba Rivers in Cameroon. There have been a few isolated reports from eastern Democratic Republic of the Congo, particularly around the Uele River east of Niangara, near the border with Sudan. There have been no reports of the animals in Gabon, but this does not mean they are not present there. The Uele River, which is the fifth largest in Africa, is a tributary of the Oubangui, and would give the animals in question open access to the Oubangui River system, which spreads out via a series of smaller rivers and tributaries linking Congo-Kinshasa with Congo-Brazzaville, Cameroon, the Central African Republic, allowing passage into Gabon.

Identity

The presence of a large aquatic or amphibious horned animal in Africa seems at first to be unlikely. Although the Indian rhino is a very good swimmer, and is quite at home in ponds and muddy pools, none of the known African rhinos can swim at all. This does not necessarily exclude the presence of a rhino that has adapted to a semi-aquatic lifestyle, but there are some interesting features that make our horned mystery stand out. The natives claim that the actual horn is said to be of polished ivory. Rhinos possess horns made from keratin, a fibrous material akin to hair. Such adornments are easily torn off whenever rhinos engage other animals in combat, such as elephants and hippos. The rhinoceros almost always loses the battle, unlike our ferocious but mysterious aquatic animal that kills elephants, hippos, and buffalos with apparent ease. Some physical features of the animal seem to have become confused with other known and unknown animals, such as the presence of a long heavy tail or a horse-like mane. According to other eyewitnesses that I have interviewed over the years, the animals possessed thin tails, akin to that of an elephant. Further research in Cameroon not only supports the presence of the animals in the

William Rebsamen

rivers there, but recent reports of the animals being killed in elephant traps and shot by hunters suggest that they exist and are still present in at least three Congo Basin countries.

In 1981, the African-American explorer Herman Regusters was told by his Boha forest guides of a recent discovery of three dead elephants on the Likouala-aux-Herbes River, 3 km northeast and upstream from their village. The dead animals were found close to the river's edge but with their tusks still attached, thus eliminating poachers as the cause of their demise. The Boha hunters examined the remains of the animals and found two large puncture marks in each of the elephants' abdomens, leading them to speculate that these elephants, including two healthy young bulls, had been killed by an *emela-ntouka* only a day or two prior. This river or semi-aquatic animal of the Congo (last reported in 1992) is greatly feared by the natives, even more so than the *mokele-mbembe*. The single ivory horn of the *emela-ntouka* is intriguing, though with the animal in Cameroon, the *n'goubou*, sometimes sporting two horns, this may indicate sexual dimorphism within the species, whatever that species might be.

DIET

The diet of both *emela-ntouka* and the *n'goubou* is said to be strictly vegetarian, with both animals browsing extensively on a variety of leaves and foliage, which is consistent with the rhinoceros. The double puncture wounds received by each of the three elephants would be consistent with a side-by-side horned configuration similar to that found in the fossil mammal *Arsinoitherium*. The animals present in Cameroon, however, have added more information to our growing dossier of this intriguing mystery that may allow us to make a tentative identity.

A LIVING FOSSIL?

In 1996, Timbo Robert, the Baka chief of Welele in Cameroon, led a hunting party to an elephant trail that led out of the Boumba River and into the forest to the east. Along the trail, Timbo and his team had dug an elephant trap, which was basically a large pit with sharpened staves or spikes driven into the bottom of the hole. Once the trap had been prepared, the Baka covered the trap's opening with large branches and retreated in wait of an unwary pachyderm. The following morning, Timbo and his party were astonished to find that their trap had indeed ensnared and killed a very large animal, but it was a *n'goubou*, not an elephant. The animal was described as being almost as big as a forest elephant, with hairless, smooth brownish-grey skin somewhat similar to an elephant. The four powerful limbs had feet similar to an elephant, and the tail resembled that of a rhino. The feature that stood out was the pair of huge nasal horns adorning much of the animal's head and snout.

Four years later, in November 2000, when Dave Woetzel and I made our first excursion into Cameroon, we showed Timbo and several other pygmy and Bantu groups pictures of various animals, both living and extinct. The illustration that these groups picked out as most like the *n'goubou* was *Arsinoitherium*, an extinct horned animal which possessed two huge horns fused to the nasal bones in a side-by-side configuration rather than front-to-back as with some modern rhinos. This came as a surprise to us, as the animals reported in the Congo Republic were described as possessing a single horn only. We pressed the Baka more closely on the anatomical details of the animal, but they were absolutely adamant, as were other eyewitnesses, that the *n'goubou* possesses two very prominent side-by-side horns rather than the single horn of the *emela-ntouka* of the Congo.

Although *Arsinoitherium* bore a passing resemblance to a rhino, it was in fact a paenungulate mammal related to elephants. This impressive herbivore is thought to have inhabited tropical rainforest and swamps. When alive, it stood between five and seven feet at the shoulder and was about ten feet long. The forest elephant (*Loxodonta cyclotis*) that currently inhabits the forests of the Congo Basin countries can reach up to eight feet in height in rare cases, which gives us an accurate comparison when considering the eyewitness accounts of the animal as being "as large as an elephant." The skeleton of *Arsinoitherium* is fairly robust and suggests that the beast may have been able to run much like a modern elephant or rhinoceros. It would appear that the *n'goubou* is largely immune to predation, not surprising if it is similar to *Arsinoitherium*, although large crocodiles could prey on the young and infirm from time to time. The double horns of *Arsinoitherium* were not "polished ivory," but were hollow horns growing over cores of solid bone.

The fossils of at least three species of *Arsinoitherium* have been found in Africa. These include *Arsinoitherium andrewsii*, *Arsinoitherium giganteum*, and *Arsinoitherium zitteli*. Angola, where *Arsiinoitherium zitteli* fossils have been found, shares its border with Congo-Brazzaville, Congo-Kinshasa, and Zambia, where our mystery horned animal has been reported over the past 100 years. The most recent reports, from the 1990s and into the early 21st century, indicate that the animals are still present in the two Congos, as well as Cameroon and possibly Gabon. A thorough search of Lake Bangweulu, the Luapula River, and adjacent swamps all the way to the Congo River, may yet produce excellent results.

WILLIAM REBSAMEN

Savannah N'Goubou

Perhaps the most thought-provoking report of any mystery animal that allegedly lives in Cameroon concerns what might possibly be a living ceratopsian dinosaur. The savannah or terrestrial *n'goubou* is described as a powerfully built animal approximately the size of an elephant, with between three and six horns protruding from a prominent neck frill. Other physical characteristics include a beaked mouth and a long, thin tail similar to an elephant. The animal is also described as being protected by brownish-grey armored skin. It is said to attack elephants with impunity, most likely due to territorial disputes. When questioned about its leg structure, witnesses described the limbs as very much like an elephant, and the tail is also said to be very elephant-like.

When browsing through our series of illustrations, the Baka became quite excited as they pointed to an illustration of the *Triceratops* as the closest likeness to the *n'goubou*. While they respect and fear creatures like the *la'kela-bembe* and *n'goubou*, they do not regard them as any stranger than an elephant or gorilla. Puzzled by this identification, we pressed the Baka more closely on the animal. Although the *Triceratops* was very similar to the savannah *n'goubou*, they stated the nose possessed a single prominent horn, but with between three and six horns adorning the neck frill. This feature was confirmed by other witnesses independently, including one plantation owner who drew a rough sketch of the savannah *n'goubou* for John Kirk on a piece of paper. The drawing bore a striking resemblance to a ceratopsian, with a prominent frill, six horns or spikes, and a beaked mouth. After returning to Vancouver, John compared the drawing to a *Styracosaurus*, an extinct ceratopsian dinosaur that supposedly went the way of the dodo eons ago. While the comparison is remarkable and has been repeated by close to a dozen eyewitnesses independently, ceratopsian fossils have yet to be found in Africa. This does not mean, however, that ceratopsian dinosaurs never lived there, but the lack of such fossils on that continent does present a stumbling block to the theory, as do other factors, including its offspring, structure of the limbs, food supply, and general habitat.

Most, if not all, ceratopsian dinosaurs are thought to have laid eggs. The savannah *n'goubou*, however, is reported to give birth to live young. According to experienced hunters who observed these animals deep in the bush, the animals move around in small herds of up to six individuals, and often conceal themselves in the forest, possibly when foraging for food. The female *n'goubou* is said to give birth to a single live calf, which it then rears alone away from the herd.

These animals are herbivorous, browsing on grasses, leaves, shoots, and other foliage, as do elephants and rhinos. An *n'goubou* approximately the size of a forest elephant would have to eat between 300 to 600 pounds of food daily. If the two animals are sharing the same habitat, and more or less the same food supply, then native reports of heated, ongoing battles between the two should not be surprising.

In November 2000, Pierre Sima visited the village of Ndelele near the border of the Central African Republic. Two days before his arrival, village hunters had shot an *n'goubou*, which they quickly butchered for its meat. The horns has been sawn off and sold to a French employee of a logging company, while the last few remains and bones had been given to the village dogs to finish off.

Although Pierre did not have our binder with the various illustrations to show to the villagers, he was able to establish that the animal that had been slain was not a rhinoceros, but an *n'goubou* with a heavy neck frill, several prominent horns or spikes, and a beaked mouth. He was even invited to dine with the villagers, and commented later that the animal's meat tasted like pork!

The question is, what kind of animal could the *n'goubou* be? The pygmy drawings of the animal reminds one of a rhino, except for the armored body and neck frill. But what do we make of the beaked mouth? I would venture a guess and suggest that the beaked mouth may be nothing more than an exaggerated prehensile lip, somewhat similar to that of the black rhino. However, until an actual specimen is made available to study, the elephant-killer of the savannah remains an intriguing mystery.

N'Goubou: Ethnological Variants

LOCATION	TRIBE	NAME OF ANIMAL
Cameroon – East Province	BaBinga tribe	N'Goubou (Savannah)
Cameroon – East Province	BaBinga tribe	N'Goubou (River)
D.R. Congo (Kinshasa)	BaKongo Sub-group	Chipekwe
P.R. Congo (Brazzaville)	Bakongo Sub-group	Emela-Ntouka
P.R. Congo (Brazzaville)	Bomitaba	Emia N'touka
P.R. Congo (Brazzaville)	Bakongo Sub-group	Aseke Moko
Gabon	Fang	Unknown
Zambia	Wa-Ushi	Chimbere

> It takes many Baka with spears to kill a Yoli.
>
> —Dondolo, Baka Elder

13
YOLI—THE SNAKE DRAGON

Perhaps the most unusual and thought-provoking of all the odd and mysterious animals my colleagues and I have heard about is the *yoli*, or snake dragon. The *yoli* is described as a giant snake, much bigger than a python, but possessing two small front limbs, which it uses to move around on land or to gain purchase when climbing trees in search of its food. The animal devours birds and monkeys, but also kills and consumes other prey, including unwary humans! The coloration of the *yoli* is somewhat similar to that of a python, with a "crown" or comb-like appendage adorning its head. This strange creature is also said to be able to mimic the sounds of certain forest animals in order to draw them to their deaths. Perhaps the most intriguing physical feature attributed to this strange creature is its alleged ability to generate an electrical charge from a horn-like appendage on its tail, similar in appearance to a rattlesnake's rattle. Everyone we interviewed about the *yoli* stated that the creatures are definitely reptilian and are extremely dangerous. However, the occasional *yoli* has been killed by Baka hunting parties, as is their habit whenever they encounter snakes and other kinds of reptiles. Sometime in mid-2000, Pierre Sima was taken to a spot in the forest, east of the Boumba River by a group of Baka hunters where they claimed a *yoli* had rested the night before. To his amazement, Pierre measured out a flattened area of about five meters square, which included small trees that had been bent over as the huge animal coiled up for the night.

Although the *yoli* is often described as a giant snake with the presence of legs, the nearest "mystery animal," or cryptid, that is comparable is the *nguma-monene*, or "giant snake" in the Lingala language of the Congo Republic. The Aka pygmies who live along the Motaba River, a tributary of the mighty Oubangui, are the most familiar with this animal and describe it as an enormous snake-like reptile with a serrated ridge running the length of the animal's body, ending in a large snake-like head with a forked tongue that darts in and out, also like a snake. However, the *nguma-monene* is far longer than the largest reported pythons, reaching a purported

WILLIAM REBSAMEN

length of between 130 to 190 feet. Almost all the eyewitnesses reporting this animal have been natives, and all of them have been emphatic that *nguma-monene* is not just a large snake, but is quite different.

We know of at least one western eyewitness who encountered an *nguma-monene* and was able to corroborate the native stories about the creature. In November 1971, on a clear, sunny day, Pastor Joseph Ellis was heading north on the Motaba River in his 30-foot-long motorized canoe. He had just dropped off a native companion and was continuing on his way when an amazing sight immediately arrested his attention. Moving just across the river from the right bank, at a distance of 100 feet, was an elongated snake-like creature with a series of diamond-like ridges running the length of its back.

Two thoughts immediately raced through Ellis' mind: "It's as long as my dugout, and it's got a back like a saw!" Ellis shut off his engine and watched as the monster swam slowly across the swollen river, which was about 15 to 20 feet deep at the time, owing to the fact that it was the height of the rainy season. The animal exited the river on the left bank, crawled through the tall grass, and headed into the jungle. At no time did Ellis observe the head of the animal, but its visible length was at least 30 feet, and it was grayish-brown in color. At first Ellis thought about following the animal into the bush in order to get a closer look, but quickly reconsidered as the creature might be dangerous.

Shortly after his amazing encounter, the missionary reached Mataba village where he was scheduled to hold a Bible study class. Still somewhat shaken by his experience, he tried to warn the villagers about the monster in the river that he had encountered, but was even more shocked to find that no one seemed to care, or at least they did not want to hear about it. Sensing that the subject might be taboo, he dropped the matter and went about the Lord's business.

I had the pleasure of speaking with Pastor Ellis in 1987 over the telephone and discussing his encounter. He was most courteous and answered my questions patiently, explaining in detail his encounter with the animal sixteen years before. In 2004, when writing my second book, *Missionaries and Monsters*, I again spoke with Pastor Ellis, who recounted the same details with unwavering accuracy and without any additional embellishments. He even provided a portrait photograph of himself and his wife during their service in the Congo. In a 1986 letter to me regarding his encounter, he wrote: "I had no idea that such a monster existed. It was shocking to see something as long as my boat swimming across the river in front of me."

Pastor Ellis was an excellent eyewitness for two reasons. First, he was not in the least interested in extant "dinosaurs" still stalking the wilds of Africa, nor had he even heard of any strange animals that allegedly existed in the swamps and jungles

William Rebsamen

Pastor and Mrs. Joseph Ellis

of the Congo. Secondly he was a missionary with only one goal in mind, and that was to preach the Gospel to the native inhabitants of the Likouala region. In his ten years of service in the Congo from 1963 to 1973, Pastor Ellis had seen many different animals, including crocodiles, monitor lizards, pythons, turtles, hippos, elephants, and marauding leopards. But the monster of the river he encountered on that fateful day in 1971 was none of these, he was absolutely certain.

One other general reference to a gigantic snake-like creature in Africa concerns the information collected by H. C. Jackson, a former British Deputy Governor, who in 1923 published a study of the Nuer people of the Upper Nile Province. The Nuer live in the marshy and savannah country on both banks of the Nile River in southern Sudan. They describe a creature called the *lau*, a large snake-like animal of monstrous proportions, much larger than a python in both length and girth, but different in color, with a bone or horn-like protrusion at the tail and a short crest of hair at the back of the head. The *lau* was said to be between 40 to 100 feet in length, its coloration ranging from brown to dark yellow, and inhabiting the great swamps of the Nile valley. It was greatly feared by the people.

The Dinka people, on the other hand, are one of the largest and most powerful ethnic groups in Southern Sudan. The Dinka speak five major dialect divisions, but the name Jaang is a general term covering these different languages. The *lau* spoken of by the Dinka people more closely resembles the *mokele-mbembe* of the Congo, with at least one report of the animal involved in the capsizing of a local chief's canoe in the Sobat river, a tributary of the White Nile, in 1992.

Although we can be fairly certain that the name *lau* refers to two distinctly different animals, both suspected to be giant unknown reptiles of some kind, by two different tribal groups speaking related Nilo-Saharan languages, another possibility may be that the name *lau* is merely a generic term used to refer to any strange or mysterious animal that both tribal groups fear. The mention of tentacles that the *lau* possesses may be honest confusion between the giant, snake-like *lau* and a rare outsized catfish that could well exist in the vast river system and swamps of southern Sudan. The question remains, could the *yoli* of Cameroon be the same as the *nguma-monene* of the Congo and the *lau* of Sudan? It would appear on the surface that they are, as the creatures share several physical features. However, legs or appendages have not been ascribed to the *lau* by the Nuer people, nor have there been any mentions made of an ability to generate electricity. The only vocalizations attributed to the *lau* is a deep rumbling or booming sound, which is quite different from the multiple vocalizations claimed for the *yoli* by the Baka in Cameroon.

One other reference to a similar creature comes from our friend Lucien Blancou, the former French game warden who worked in the Likouala Region of the now

Republic of the Congo. In 1945, his gun bearer, a Congolese native called Mitikata, told Blancou about a huge serpent with legs called the *ngakouala-ngou*, that frequented the rivers and swamps. Mitikata examined the alleged tracks of an *ngakouala-ngou* himself in 1930 near an outpost at Ndele in former Ubangui-Shari, now known as the Central African Republic. The name *ngakouala-ngou* comes from the Bolgo language, a member of the Bua language group, spoken only by about 1800 people today in south-central Chad.

With at least four different (known) references to what is possibly the same elongated, snake-like reptile within four different Congo Basin countries by at least six different tribal groups, it seems a fair bet that we are dealing with one and the same animal. The puzzle, however, is not yet complete, as many of the pieces are still missing, particularly concerning details of the animal's reproduction and ability to generate a mild electrical charge. Like almost all reptiles, the *yoli* is said to lay a number of eggs. The Baka claim that these eventually hatch into different kinds of colorful snakes, with only a few of the offspring actually being *yoli*.

Significant electrical production (electrogenesis) in biological organisms is known primarily from fish (electric eels, electric rays, electric catfish, etc.), with unconfirmed allegations of one electric mammal, the giant otter shrew (*Potamogale velox*) of central Africa. The electric eel (*Electrophorus electricus*) is actually a cylindrical South American knifefish that grows to six feet in length, and produces a dangerous shock up to 500 volts. Electric catfish include nineteen species in the genera *Malapterurus* and *Paradoxoglanis*. They are found throughout western and central Africa, and the Nile River. One suggestion is that the *yoli* may be a giant electric fish that can comfortably spend time on dry land for a time. Some species of catfish are able to survive for periods out of water, so long as they keep moist. But, if our mystery animal does turn out to be a giant semi-aquatic reptile that can emit an electrical impulse, then it will be unique in every respect.

One final intriguing piece of information concerns the colorful offspring of the *yoli*, of which only a few are said to be actual *yolis*. As it is unlikely that any hunting party, whether Baka or Bantu, would sit around watching eggs hatch in the presence of a monstrous, protective female reptile, this may simply be an honest mistake on the part of our native informants. As *yolis* have been killed from time to time by spear-toting pygmy hunters, it is surely only a matter of time before a juvenile is captured or killed and made available to study. This will bring us much closer to determining if this creature is a new species, and perhaps even one and the same as the *lau*, the *ngakouala-ngou*, and the *nguma-monene*.

YOLI: ETHNOLOGICAL VARIANTS

LOCATION	TRIBE	NAME OF ANIMAL
Cameroon – East Province	BaBinga tribe	Yoli
Central African Republic	Mboum Tribe	Ngakouala-Ngou
P.R. Congo (Brazzaville)	Bakongo Sub-group	Nguma-Monene
Sudan	Nuer Tribe	Lau

William Rebsamen

William Rebsamen

Appendix I

Lingala Words and Phrases

Hello!	Losako!
Good day.	Mbote.
Good night.	Butu elamu.
How are you?	Sango nini?
Fine, thanks.	Malamu, melesi.
Goodbye.	Kendeke malamu.
See you later!	Tokomonono na nsima!
Where are you going?	Limbisa ngai?
Yes.	Iyo.
No.	Te.
Please.	Palado.
Thank you.	Natondi yo.
You're welcome.	Likambote.
I understand.	Nayoki.
I don't understand.	Nayoki te.
How much?	Boni?
Book	Buku
Knife	Mbeli
Paper or Document	Mokanda
Kitchen	Kuku
Water	Mai
Milk	Mabele
Coffee	Kawa
Tea	Ti
Coffee with milk	Kawa na mabele
Potato	Libenge
Egg	Likei

Rice	Loso
Orange	Lilala
Animal (generic), or Meat	Nyama
Snake	Nyoka
Frog	Mombemba
Fish	Mbisi
Elephant	Nzoku
Hippo	Ngubu
Crocodile	Nkoli
Leopard	Koi
Woman	Mwasi
Man	Mobali
Family	Libota
Father	Tata
Mother	Mama
Child	Mwana
White man	Mondele
Ear	Litoi
Eye, Eyes	Liso, Miso
Tooth, Teeth	Lino, Mino
Mouth	Monoko
Nose	Zolo
Foot (Leg), Feet (Legs)	Lokolo, Makolo
Hand, Hands	Loboko, Maboko
Tree	Nzete
Tall Tree	Nzete molai
Strong Tree	Nzete makasi
Flower	Fulele
Forest	Zamba
Lake	Etima
Sky	Likolo
Sand	Zelo
Street	Balabala or Mololo
Day	Mokolo
Year	Mbula
Month	Sanza
Hour	Ngonga
What time is it?	Nogongaboni?

Appendix II

Baka Words and Phrases
(From Field Notes by Bill Gibbons and Peter Beach)

Table	Mesa
Chair	Bunga
Sit down	Ma titi
Beard	Lukaka
Eyebrows	Jinja
Head hair	Singu
Finger	Lebabo
Fingernail	Cuquebabo
Nose	Banga
Nose hair	Banga bo
Ears	Jeb
Throat	Limbe
Cheeks	Limbuto
Lips	Mobo
Teeth	Tebo
Tongue	Melbo
Back of throat	Bako
Gray hair or straight hair	Busa
Shoulders	Sakose
Collar bone	Geebee
Neck	Ngabo
Heart	Tobo
Side	Gabo
Forearm	Babo
Arm bone	Monono
Elbow	Aaoo
Dark skin	Quoto bebe

Light skin	Quoto boba
Big toe	Qukwanoli
Body hair	Susu
Belly button	Munyabo
Belly	Mmjooubo
Under finger nail	Bendu
Fold arm	Bayatem
Headache	Lucala
Tired or resting	Mawala
A cross look with the eyebrows	Macoco jimja
See or sight	Masea
Write	Maade
Walk	Waloo
Run	Moulenya
To refuse	Makuway
Go back	Tolole
Baby	Makuba
Female	Laca
God	Do gad beay
Village	Billi
Sky	Leblo
Sun	Sungu
Trees	Ngwalu
Forest	Forrey
Antibiotic plant	Musimbi
Thorns	Titi
Little leaf	Yegali
Big leaf	Makamba
Cut leaves	Momu
Dirt	Tulu
Ground litter	Mushumbu
Edge	Taan
Coming from forest	Mabe ata
Net	Juka
Catch with net	Tindele
Sack	Sackosee
Pick up sack	Bebe
Brush	Wulu

Strap	Ku
Stuff inside bag	Posee
Handle of machete	Limba
Handkerchief	Masui
Yellow	Jene
Blue,	Naki naki
Zipper	Maka
Close zipper	Maka kumsa
Sit down on ground	Titi kumsa
Lay down on ground	Lati kumsa
Lay prone	Wee kumsa
Big or great	J'ba
Dragonfly	Tindili
Flies	Jimji
Spider	Fofi
Giant Spider	Jiba fofi
Turtle	Kunda
Snake	Pikolo
Elephant	Ya'a
Hippo, Rhino, Two-horned Cryptid	N'goubou (a tonal word)
Gorilla	Ebobo
Chimpanzee	Seko
Monkey	Kema
White man	Mboungue

WILLIAM REBSAMEN

APPENDIX III

ETHNOLOGICAL INDEX OF MYSTERY ANIMAL NAMES

Amali   19, 50, 210

Aseke moko   239, 247

Bagidui   33-34, 210, 224

Behemoth   11-14

Chimbere   247

Chipekwe   24, 28, 235, 239-240, 247

Diba   33, 210, 224

Dodu   118, 137-138, 141, 159

Embulu-embembe   211, 224

Emela-ntouka   95, 97, 106, 236, 238-239, 242-243, 247

Emia n'touka   239, 247

Esamba   76-78

Isiququmadevu   24, 210, 224

Jago-nini   19, 50, 103, 211, 224

Jba fofi   158-159

Kalinoro   111

Kowuru   234

La'kila-bembe   110, 114-125, 133-137, 140-142, 159-160, 162-163, 165-168, 170-173, 175, 180-184, 188-189, 197-200, 212, 215-219, 224

Lau   210, 224, 254-256

Lukwata   210, 224

Mbilintu   24, 210, 224

Mbokalemuembe, or *M'bokale-muembe*   36, 211, 224

M'koo-m'bemboo   35, 50, 198, 211, 224

Mokele-mbembe   27, 36-7, 40-41, 43-44, 47-50, 52-58, 62-63, 67, 69-75, 79-85, 94-99, 103, 106, 109-110, 115, 136-137, 150, 160-161, 167, 175, 181, 189, 191-192, 198-199, 205-224, 238, 254

Morou-ngou 210
Ngakouala-ngou 255-256
N'goubou 124, 141, 159, 233-247
Nguma-monene 249-256
Nwe 38, 41, 43, 224
N'yamala 48-50, 109, 210, 224
P'ih-mw 13
Sirrush 12-13
Songo 33, 210, 224
Yoli 249-256

BIBLIOGRAPHY

Agnagna, Marcellin. 1983. Report on the Results of the First Congolese Mokele-mbembe Expedition, 1983. *Cryptozoology* (2): 103-112.

Akowuah, Thomas A. 1996. *Lingala-English, English-Lingala Dictionary and Phrasebook*. New York: Hippocrene Books.

Bright, Michael. 1984. Meet Mokele-mbembe. *BBC Wildlife Magazine* (12): 596-601.

Conan Doyle, Arthur. 1912. *The Lost World*. London: Hodder & Stoughton.

Hagenbeck, Carl. 1912. *Beasts and Men*. New York: Longmans, Green & Co.

Heuvelmans, Bernard. 1959. *On the Track of Unknown Animals*. New York: Hill & Wang.

Heuvelmans, Bernard. 1978. *Les Derniers Dragons d'Afrique*. Paris: Plon.

Hichens, William. 1937. African Mystery Beasts. *Discovery* (December): 369-373.

Horn, Aloysius Smith. 1927. *The Ivory Coast in the Earlies*. New York: Simon & Shuster.

Hughes, J. E. 1933. *Eighteen Years on Lake Bangweulu*. London: The Field.

International Rhino Foundation. 2010. website. http://www.rhinos-irf.org/

Jackson, H. C. 1923. *The Nuer of the Upper Nile Province*. Khartoum, Sudan: El Hadara.

Johnston, Henry. 1906. *Liberia*. Vol. 2. London: Hutchinson and Co.

Ley, Willie. 1948. *The Lungfish, The Dodo, and the Unicorn*. New York: Viking Press.

MacDonald, Fiona, et al. 2000. *Peoples of Africa*. Vol. 3. New York: Marshall Cavendish.

Mackal, Roy P. 1980. *Searching for Hidden Animals*. Garden City, NJ: Doubleday.

Mackal, Roy P. 1987. *A Living Dinosaur? In Search of Mokele-mbembe*. Leiden: E. J. Brill.

Odhner, John D. 1981. *English-Lingala Manual*. Lanham, MD: University Press of Ameria.

Pinkerton, John. 1814. *A General Collection of the Best and Most Interesting Voyages and Travels in all Parts of the World.* Vol. 16. London: Longman, Hurst, Rees, and Orme.

Proyart, Abbé Lievain Bonaventure. 1776. *Histoire de Loango, Kakongo, et Autres, Royaumes d'Afrique; Rédigée d'après les Mémoires de Préfets Apostoliques de la Mission françoise.* Paris: C. P. Berton & N. Crapart.

Regusters, Herman A. 1982. *Mokele-Mbembe: an investigation into rumors concerning a strange animal in the Republic of the Congo, 1981.* Munger Africana Library Notes, 64. (http://caltechmaln.library.caltech.edu/71/)

Sanderson, Ivan T. 1948. There Could be Dinosaurs. *Saturday Evening Post* (Jan. 3): 17, 53-56.

Sanderson, Ivan T. 1969. *More Things*. New York: Pyramid Books.

Schomburgk, Hans. 1910. *Wild und Wilde im Herzen Afrikas*. Berlin: Egon Fleishel & Co.

Shuker, Karl P. N. 1989. *Mystery Cats of the World*. London: Robert Hale.

Shuker, Karl P. N. 1995. *In Search of Prehistoric Survivors: Do Giant 'Extinct' Creatures Still Exist?* London: Blandford Press.

Summers, Roger. 1959. *Prehistoric Rock Art of the Federation of Rhodesia and Nyasaland*. Rhodesia and Nyasaland: National Publications Trust.

Trial, Georges. 1955. *Dix Ans De Chasse Au Gabon*. Paris: Crépin-Leblond & Cie.

In Memoriam

Tim Dinsdale   (1924 – 1987)
Bernard Heuvelmans   (1916 – 2001)
Herman Regusters   (1933 – 2005)
Eugene P. Thomas   (1925 – 2005)
J. Richard Greenwell (1942 – 2005)
Phillip Anderton   (1957 – 2005)
Scott T. Norman   (1965 – 2008)
Marie T. Womack   (*n. d.* – 1992)

Coachwhip Publications

CoachwhipBooks.com

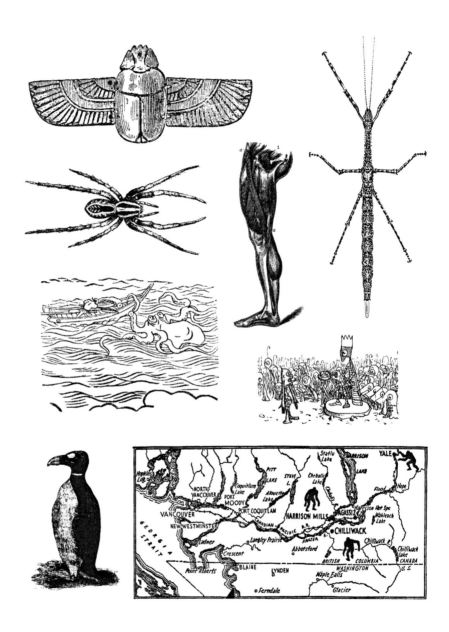

COACHWHIP PUBLICATIONS
CoachwhipBooks.com

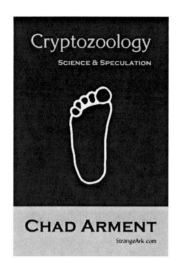

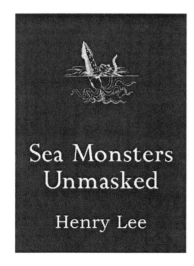
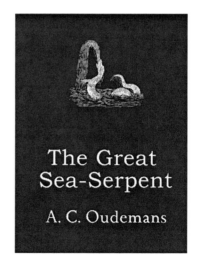

Cryptozoology, Natural History, and More...

CPSIA information can be obtained at www.ICGtesting.com
Printed in the USA
BVOW080915071211
277790BV00003B/174/P